ATLAS OF INTRODUCTORY MYCOLOGY

Second Edition

Richard T. Hanlin
Department of Plant Pathology
University of Georgia
Athens, Georgia

Miguel Ulloa
Departamento de Botanica
Instituto de Biologia
Universidad Nacional Autonoma de Mexico
Mexico 20, D.F., Mexico

Hunter Textbooks Inc.

823 Reynolda Road
Winston-Salem, North Carolina 27104

Copyright 1988 by Richard T. Hanlin and Miguel Ulloa

Printed in the United States of America

ISBN 0-88725-092-0

Library of Congress Catalog Card Number 78-65422

All rights reserved. No part of this book may be reproduced in any form without permission in writing, except by a reviewer, who may quote brief passages in critical articles and reviews.

Preface

Any work of this kind necessarily includes contributions from many individuals besides the authors. We are grateful to our students who have offered valuable suggestions and tested the work sheets that preceded the Atlas. We also appreciate the comments and assistance of various colleagues as the Atlas was developed.

The approach used in the Atlas is strongly morphological and supports our philosophy that a solid knowledge of fungus structure and life history is essential to whatever else one may wish to do with fungi. The Atlas is designed to help the student learn about fungi. If it succeeds in this, then it will have fulfilled its purpose.

Introduction to Second Edition

Since publication of the first edition of the Atlas there has been a steady flow of new information in the literature, resulting in reinterpretations of data and changes in concepts. Some of these new ideas have been incorporated into this revision to keep it as current as possible, consistent with the authors' own taxonomic philosophies. Changes in nomenclature have been made to conform to the recent taxonomic literature.

In basic approach and concept the Atlas remains the same, with emphasis on morphology and life histories of fungi. Some additional genera have been added to provide instructors with more flexibility in choosing materials to use in their own laboratories. The general arrangement continues to conform most closely to that of *Introductory Mycology*, 3rd edition, by C.J. Alexopoulos and C.W. Mims, but the simplified taxonomic scheme of the first edition has been retained. References have been updated and in some areas expanded.

We thank Melvin S. Fuller for reviewing this revision and providing helpful suggestions.

We are grateful for the response the Atlas has received and hope that it will continue to provide an aid to introducing students to the fascinating world of fungi.

Richard T. Hanlin
Miguel Ulloa

Athens, Georgia
May 1988

Introduction to First Edition

The idea for an atlas of introductory mycology evolved as a result of observing students in laboratory over a period of several years. Our experience indicates that learning is greatly reduced because students spend too much time searching for the various structures they are to study, and that there are few references they can consult to aid them. It was felt that a study guide with drawings and photographs would acquaint the student with what he or she should see under the microscope and thus permit him or her to locate and study these structures more efficiently. An attempt has been made to show various structures as the student will see them. For this reason only a few photographs were made under oil and phase contrast since we feel these are seldom used in an introductory laboratory. Likewise, materials have been selected that are readily grown or obtained by the instructor and which show features pertinent to gaining a basic knowledge of fungal morphology.

The atlas is designed to be used in a one-quarter or semester course which emphasizes the morphology and taxonomy of fungi. In general approach it most resembles that of *Introductory Mycology*, 2nd edition, by C.J. Alexopoulos. The taxonomic scheme conforms most closely to that given in volumes IVA and IVB of *The Fungi – An Advanced Treatise*, edited by G.C. Ainsworth, F.K. Sparrow, and A.S. Sussman, but some changes have been made for pedagogic and personal reasons. Information on the ecology and utilization of fungi is summarized in *Fundamentals of the Fungi*, by E. Moore-Landecker. A list of other introductory level references is provided elsewhere in the atlas.

This atlas is intended as an aid, not a replacement, for examining materials in the laboratory. Whenever possible we use living materials as students find them more interesting. Experience in working with fungi is provided through development of a culture collection.

The references listed are not intended to be comprehensive but are selected as sources of information for use by the student while studying the laboratory materials. More extensive lists of references can be found in the general references given. Because most students lack facility with foreign languages, most of the references are in English.

The genera listed in the "Outline of Classification of Fungi Studied in Introductory Mycology" includes all genera mentioned, even incidentally, during the course; consequently not all of them will be found in the text.

All drawings are originals made for the atlas by Dr. Miguel Ulloa. Unless otherwise indicated, the photographs are originals taken by the authors.

We are grateful to Sue Glover, M.S., for typing the manuscript, and to Dr. John P. Jones for assistance in printing most of the photographs. Without their diligence the Atlas would not have been completed on time. Lastly, we wish to express our gratitude to officials of the Universidad Nacional Autonoma de Mexico for granting sabbatical leave to Miguel Ulloa during the 1975-76 academic year so we could complete the illustrations for the Atlas. We are also grateful to the Consejo Nacional de Ciencia y Tecnologia (CONACyT) of Mexico for providing him with financial assistance during his leave.

Richard T. Hanlin	Athens, Georgia
Miguel Ulloa	August, 1978

TABLE OF CONTENTS

Preface	iii
Introduction to Second Edition	iv
Introduction to First Edition	v
Outline of Classification of Fungi Studied in Introductory Mycology	ix
Isolation of Fungi from Natural Substrates	2
Isolation of Acrasiomycetes	3
Isolation of Myxomycetes	3
Isolation of Zoosporic Aquatic Fungi	4
Isolation of Zygomycetes	6
Isolation of Ascomycetes and Deuteromycetes	6
Isolation of Basidiomycetes	8
Composition of Media and Solutions Used in Introductory Mycology	10
Mycology Culture Collection	19
General References	21
Kingdom Fungi	22
Myxomycota	22
Acrasiomycetes	22
Protosteliales	23
Dictyosteliales	23
Myxomycetes	27
Ceratiomyxomycetidae	27
Ceratiomyxales	27
Myxogastromycetidae	27
Physarales	27
Trichiales	28
Stemonitales	28
Liceales	29
Eumycota	34
Mastigomycotina	34
Chytridiomycetes	34
Chytridiales	35
Spizellomycetales	35
Blastocladiales	36
Monoblepharidales	36
Hyphochytriomycetes	42
Hyphochytriales	42
Plasmodiophoromycetes	43
Plasmodiophorales	43
Oomycetes	44
Saprolegniales	44
Peronosporales	44
Zygomycotina	49
Zygomycetes	49
Mucorales	49
Entomophthorales	51
Trichomycetes	53
Harpellales	53
Deuteromycotina	62
Blastomycetes	62
Sporobolomycetales	62
Cryptococcales	62
Hyphomycetes	62
Moniliales	63
Coelomycetes	65
Melanconiales	65
Sphaeropsidales	65
Medical Mycology	72
Ascomycotina	74
Hemiascomycetes	74
Endomycetales	74
Taphrinales	75
Euascomycetes	79
Plectomycetidae	79
Eurotiales	79
Onygenales	80
Microascales	80
Pyrenomycetidae	83
Erysiphales	83
Chaetomiales	88
Phyllachorales	89
Sordariales	93
Xylariales	98
Diaporthales	102
Hypocreales	106
Clavicipitales	108
Discomycetidae	112

TABLE OF CONTENTS, cont'd

Pezizales	112
Helotiales	118
Ostropales	120
Phacidiales	122
Tuberales	123
Laboulbeniomycetes	124
Laboulbeniales	124
Loculoascomycetes	127
Myriangiales	127
Pleosporales	130
Dothideales	132
Hysteriales	133
Basidiomycotina	136
Heterobasidiomycetes	136
Uredinales	136
Ustilaginales	138
Tremellales	140
Auriculariales	140
Septobasidiales	141
Holobasidiomycetes	142
Hymenomycetidae	142
Dacrymycetales	142
Exobasidiales	143
Agaricales	144
Aphyllophorales	154
Gasteromycetidae	166
Lycoperdales	166
Sclerodermatales	167
Tulostomatales	167
Nidulariales	167
Phallales	168
Mycorrhizae	172
Lichens	174
Fossil Fungi	178
Fungal Ecology	179
Fungal Physiology	184
Fungi of Commerical Value	188
Mycotoxins	189
Index	191

Outline of Classification of Fungi Studied in Introductory Mycology

Kingdom Fungi
DIVISION MYXOMYCOTA

Class	Subclass	Order	Family	Genus
Acrasiomycetes	Protostelidae	Protosteliales	Protosteliaceae	*Protostelium*
	Dictyostelidae	Dictyosteliales	Dictyosteliaceae	*Dictyostelium* *Polysphondylium*
Myxomycetes	Ceratiomyxomycetidae	Ceratiomyxales	Ceratiomyxaceae	*Ceratiomyxa*
	Myxogastromycetidae	Physarales	Physaraceae	*Fuligo* *Physarum*
		Trichiales	Trichiaceae	*Arcyria* *Hemitrichia*
		Stemonitales	Stemonitaceae	*Comatricha* *Diachea* *Stemonitis*
		Liceales	Reticulariaceae	*Lycogala*
			Cribrariaceae	*Dictydium*

DIVISION EUMYCOTA

Subdivision Mastigomycotina

Class	Subclass	Order	Family	Genus
Chytridiomycetes		Chytridiales	Synchytriaceae	*Synchytrium*
			Chytridiaceae	*Allochytridium* *Cylindrochytridium* *Rhizophydium*
		Spizellomycetales	Spizellomycetaceae	*Rhizophlyctis*
			Olpidiaceae	*Olpidium* *Rozella*
		Blastocladiales	Blastocladiaceae	*Allomyces* *Blastocladiella*
		Monoblepharidales	Monoblepharidaceae	*Monoblepharis*
Hyphochytriomycetes		Hyphochytriales	Rhizidiomycetaceae	*Rhizidiomyces*
Plasmodiophoromycetes		Plasmodiophorales	Plasmodiophoraceae	*Plasmodiophora* *Spongospora*
Oomycetes		Saprolegniales	Saprolegniaceae	*Achlya* *Aphanomyces* *Saprolegnia*
		Peronosporales	Pythiaceae	*Phytophthora* *Pythium*
			Peronosporaceae	*Bremia* *Peronospora* *Plasmopara*

Subdivision Mastigomycotina, continued

Class	Subclass	Order	Family	Genus
			Albuginaceae	*Albugo* (= *Cystopus*)

Subdivision Zygomycotina

Class	Subclass	Order	Family	Genus
Zygomycetes		Mucorales	Mucoraceae	*Actinomucor* *Mucor* *Phycomyces* *Rhizopus* *Zygorhynchus*
			Pilobolaceae	*Pilobolus*
			Thamnidiaceae	*Chaetostylum* *Thamnidium*
			Choanephoraceae	*Blakeslea* *Choanephora*
			Cunninghamellaceae	*Cunninghamella*
			Syncephalastraceae	*Syncephalastrum*
			Kickxellaceae	*Coemansia* *Linderina*
			Endogonaceae	*Endogone* *Gigaspora* *Glomus*
		Entomophthorales	Entomophthoraceae	*Basidiobolus* *Entomophthora* (= *Empusa*)
		Zoopagales	Zoopagaceae	*Stylopage*
Trichomycetes		Eccrinales	Eccrinaceae	*Enterobryus*
		Harpellales	Genistellaceae	*Smittium*

Subdivision Deuteromycotina

Class	Subclass	Order	Family	Genus
Blastomycetes		Sporobolomycetales	Sporobolomycetaceae	*Sporobolomyces*
		Cryptococcales	Cryptococcaceae	*Candida* *Cryptococcus* *Rhodotorula*
Hyphomycetes		Moniliales	Agonomycetaceae	*Papulaspora* *Rhizoctonia* *Sclerotium*

Subdivision Deuteromycotina, continued

Class	Subclass	Order	Family	Genus
			Moniliaceae	*Arthrobotrys*
				Aspergillus
				Chromelosporium
				Epidermophyton
				Geotrichum
				Gliocladium
				Histoplasma
				Microsporum
				Monilia
				Oedocephalum
				Oidium
				Ostracoderma
				Ovulariopsis
				Penicillium
				Spiniger
				Trichoderma
				Trichophyton
				Trichothecium
				Verticillium
			Dematiaceae	*Alternaria*
				Bipolaris
				Cercospora
				Chalara
				Cladosporium
				Curvularia
				Helicosporium
				Helminthosporium
				Orbimyces
				Pithomyces
				Spilocaea
				Thielaviopsis
			Stilbellaceae (Stilbaceae)	*Dendrostilbella*
				Graphium
				Pesotum
				Stilbella
			Tuberculariaceae	*Epicoccum*
				Fusarium
				Myrothecium
				Sphacelia
				Starkeyomyces
				Tubercularia
Coelomycetes		Melanconiales	Melanconiaceae	*Colletotrichum*
				Pestalotia
		Sphaeropsidales	Sphaeropsidaceae	*Cytospora*
				Phoma
				Phomopsis
				Septoria
			Zythiaceae	*Endothiella*

Subdivision Ascomycotina

Class	Subclass	Order	Family	Genus
Hemiascomycetes		Endomycetales	Dipodascaceae	*Dipodascopsis* *Dipodascus*
			Cephaloascaceae	*Cephaloascus*
			Saccharomycetaceae	*Saccharomyces* *Saccharomycodes* *Schizosaccharomyces*
			Spermophthoraceae	*Nematospora*
		Taphrinales	Taphrinaceae	*Taphrina*
Euascomycetes	Plectomycetidae	Eurotiales	Eurotiaceae	*Emericella* *Eupenicillium* (= *Carpenteles*) *Eurotium* *Monascus* *Sartorya* *Talaromyces*
		Onygenales	Gymnoascaceae	*Gymnoascus* *Myxotrichum* *Nannizzia*
		Microascales	Ophiostomataceae	*Ceratocystis*
	Pyrenomycetidae	Erysiphales	Erysiphaceae	*Erysiphe* *Microsphaera* *Phyllactinia* *Podosphaera* *Sphaerotheca* *Uncinula*
		Chaetomiales	Chaetomiaceae	*Chaetomium*
		Sordariales	Melansporaceae	*Melanospora* *Thielavia*
			Sordariaceae	*Gelasinospora* *Neurospora* *Sordaria*
		Phyllachorales	Phyllachoraceae	*Glomerella* *Phyllachora*
		Xylariales	Xylariaceae	*Hypoxylon* *Xylaria*
		Diaporthales	Diaporthaceae	*Cryphonectria* *Diaporthe* *Endothia* *Gnomonia* *Stegophora*
		Hypocreales	Hypocreaceae	*Gibberella* *Hypocrea* *Hypomyces* *Nectria* *Neocosmospora*

Subdivision Ascomycotina, continued

Class	Subclass	Order	Family	Genus
		Clavicipitales	Clavicipitaceae	*Balansia* *Claviceps* *Cordyceps*
	Discomycetidae	Pezizales	Sarcoscyphaceae	*Sarcoscypha*
			Sarcosomataceae	*Urnula*
			Pezizaceae	*Peziza*
			Ascobolaceae	*Ascobolus*
			Pyronemataceae	*Aleuria* *Otidea* *Pyronema*
			Helvellaceae	*Gyromitra* *Helvella*
			Morchellaceae	*Morchella*
		Helotiales	Sclerotiniaceae	*Monilinia* *Sclerotinia* *Stromatinia*
			Dermateaceae	*Pseudopeziza*
			Geoglossaceae	*Geoglossum* *Spathularia*
			Leotiaceae	*Calycella* *Helotium* *Leotia*
		Ostropales	Stictidaceae	*Vibrissea*
		Phacidiales	Rhytismataceae	*Hypoderma* *Rhytisma*
		Tuberales	Tuberaceae	*Tuber*
			Elaphomycetaceae	*Elaphomyces*
		Caliciales	Cypheliaceae	*Cyphelium*
		Lecanorales	Cladoniaceae	*Cladonia*
			Graphidaceae	*Graphis*
			Lecanoraceae	*Lecanora*
			Parmeliaceae	*Hypogymnia* *Parmelia*
			Peltigeraceae	*Peltigera*
			Physciaceae	*Physcia*
			Stictaceae	*Pseudocyphellaria*

Subdivision Ascomycotina, continued

Class	Subclass	Order	Family	Genus
			Umbilicariaceae	*Umbilicaria*
			Usneaceae	*Alectoria* *Usnea*
Laboulbeniomycetes		Laboulbeniales	Laboulbeniaceae	*Laboulbenia*
Loculoascomycetes		Myriangiales	Myriangiaceae	*Elsinoe* *Myriangium*
		Pleosporales	Pleosporaceae	*Leptosphaeria* *Pleospora*
			Sporormiaceae	*Sporormiella*
			Venturiaceae	*Apiosporina* (= *Dibotryon*) *Venturia*
		Dothideales	Dothideaceae	*Guignardia* *Mycosphaerella*
			Pseudosphaeriaceae	*Leptosphaerulina*
		Hysteriales	Hysteriaceae	*Hysterium* *Hysterographium*

Subdivision Basidiomycotina

Class	Subclass	Order	Family	Genus
Heterobasidiomycetes		Uredinales	Pucciniaceae	*Gymnosporangium* *Puccinia* *Uromyces*
			Melampsoraceae	*Cronartium*
			Coleosporiaceae	*Coleosporium*
		Ustilaginales	Ustilaginaceae	*Ustilago*
			Tilletiaceae	*Aessosporon* (= *Sporobolomyces*) *Tilletia*
		Tremellales	Tremellaceae	*Exidia* *Tremella*
		Auriculariales	Auriculariaceae	*Auricularia*
		Septobasidiales	Septobasidiaceae	*Septobasidium*
Holobasidiomycetes	Hymenomycetidae	Exobasidiales	Exobasidiaceae	*Exobasidium*
		Dacrymycetales	Dacrymycetaceae	*Calocera* *Dacrymyces*

Subdivision Basidiomycotina, continued

Class	Subclass	Order	Family	Genus
		Agaricales	Boletaceae	*Boletus* *Strobilomyces*
			Russulaceae	*Lactarius* *Russula*
			Agaricaceae	*Agaricus*
			Amanitaceae	*Amanita*
			Cantharellaceae	*Cantharellus* *Craterellus*
			Coprinaceae	*Coprinus*
			Lepiotaceae	*Chlorophyllum* *Lepiota*
			Strophariaceae	*Psilocybe*
			Tricholomataceae	*Armillaria* *Armillariella* *Clitocybe* *Marasmius* *Panus* *Pleurotus*
		Aphyllophorales	Clavariaceae	*Clavaria* *Clavariadelphus* *Clavicorona* *Ramaria*
			Schizophyllaceae	*Schizophyllum*
			Thelephoraceae	*Corticium* *Peniophora* *Sparassis* *Stereum* *Thelephora*
			Hydnaceae	*Dentinum* *Echinodontium* *Hericium* *Hydnum* *Steccherinum*
			Polyporaceae	*Coriolus* *Daedalea* *Fomes* *Ganoderma* *Heterobasidion* *Irpex* *Laetiporus* *Lenzites* *Merulius* *Polyporus* *Poria* *Pycnoporus*

Subdivision Basidiomycotina, continued

Class	Subclass	Order	Family	Genus
	Gasteromycetidae	Lycoperdales	Lycoperdaceae	*Calvatia* *Lycoperdon*
			Geastraceae	*Geastrum*
		Sclerodermatales	Sclerodermataceae	*Scleroderma* *Pisolithus*
		Tulostomatales	Calostomataceae	*Calostoma*
		Nidulariales	Nidulariaceae	*Crucibulum* *Cyathus*
		Phallales	Phallaceae	*Dictyophora* *Mutinus* *Phallus*

ATLAS OF INTRODUCTORY MYCOLOGY

Isolation of Fungi From Natural Substrates

There are hundreds of techniques described for isolating various fungi from natural substrates, but many of these techniques are complicated and are unsuited for use by beginning students. Some of the more frequently used procedures for isolating common fungi are given below. Additional methods can be found in the references given or by consulting the articles on specific fungi which are provided for each group.

The composition of media referred to can be found in the next section of the Atlas. No single medium is suitable for culturing all fungi, although some media, such as V-8 juice, malt extract, or potato dextrose agars will support a wide variety of fungi. If you wish to determine the total fungus flora (mycoflora) of a given substrate, ideally several media should be used, but if this is impractical, then a good broad spectrum medium should be selected. Special media are necessary for the isolation and maintenance of certain kinds of fungi. Fungi growing under conditions of high osmotic concentration (low moisture), for example, must be isolated and grown on a malt-salt agar which more closely duplicates the conditions of the natural habitat. Environmental conditions may also be important in isolating and growing certain fungi. Thermophilic species can only be isolated and cultured at high temperatures, and such species will not be recovered unless this requirement is met. Thus, the selection of the medium and isolation conditions can be an important factor in determining what fungi are isolated.

With many natural substrates a succession of fungi will be observed. On dung, for example, the fast-growing mucoraceous fungi will usually appear first, followed by imperfect fungi. The perithecial ascomycetes then appear, and last will be the basidiomycetes. Substrates such as this must be observed for several weeks to see the various fungi as they occur.

Isolation of Acrasiomycetes

The acrasiaceous fungi occur in a variety of habitats, including soil, decaying organic matter, such as leaves or flower parts, bark, and dung. They appear to be most common in the soil and decaying leaf litter of deciduous forests, so material from such sites is recommended. Dried flowers and small, dry fruits (such as elm achenes) that are still attached to the plant have been excellent sources of acrasiaceous fungi in our laboratory. Place small amounts of soil or plant debris on the surface of agar plates. Best results are obtained if some of the agar surface is left free so the organisms can grow out. Dry plant materials can be soaked in distilled water before plating. The most commonly used media are weak hay infusion or lactose-yeast extract agars. Glucose and peptone are often substituted for the lactose and yeast extract. If hay is not available, dilute rabbit food agar works well.

Beginning 2-3 days after inoculation, plates should be examined frequently, under a dissecting microscope, for the presence of sorocarps. When found, the heads of the sorocarps can be removed with a sterile needle and transferred to a fresh agar plate. In some instances it may be necessary to add a bacterium such as *Escherichia coli* as food. Isolation and maintenance can be carried out on the same medium.

A more elaborate procedure for isolating the Acrasiales is to make a soil suspension and plate this out by streaking loopfuls on the surface of the agar. The agar may also be seeded previously with *E. coli* to assure a source of food. This technique is described in detail in Cavender & Raper (1965).

Isolation of Myxomycetes

The plasmodial slime molds are most readily found on decaying logs or leaves a few days after heavy rain. Plasmodia will sometimes be found beneath the bark of wet logs, but the fructifications of the Myxomycetes are more likely to be found. Isolations can be made from both types of materials.

Slime molds may develop in damp chambers from various substrates, but the outer bark of living trees or decaying logs is often a good source. Collect thin pieces of outer bark from trees or logs and place them in a moist chamber lined with filter paper or paper towelling. Pour water over the bark and let it soak overnight. Pour off the water the next day and leave at room temperature. After 4 or 5 days examine for plasmodia; these may be very small and are best seen with the aid of a stereomicroscope. When a plasmodium is found, transfer it to a petri dish containing corn meal agar prepared half strength. The plasmodium will likely be contaminated with fungi. Pure cultures can be made by waiting until the plasmodium crawls off of the block and then transferring it to fresh agar plates. When the culture is pure, transfer to another corn meal agar plate and feed with several sterile flakes of rolled oats.

Isolations may also be made from spores in sporangia, although this is often quite difficult. Prepare half strength corn meal agar and place a sporangium on the agar surface, break it open with a sterile needle, and spread the spores over the surface of the agar. Add 3-4 ml sterile water to the agar plate and incubate at room temperature. Spore germination should begin 12-24 hours after inoculation. Swarm cells should be visible under the medium power of the microscope. Five to six days after germination tiny plasmodia should have formed on the agar surface. When these are large enough to be seen without a lens, transfer them to fresh corn meal agar plates and handle as above.

Isolation of Zoosporic Aquatic Fungi

The zoosporic aquatic fungi are fungi which normally live in water and which reproduce asexually by means of zoospores. They can be found in lakes, ponds, streams, and in both fresh and marine water, as well as in most soils, where they inhabit the film of water surrounding the soil particles. They range in size from microscopic unicellular forms to those with a well developed mycelium.

Techniques for isolating zoosporic aquatic fungi vary according to the habitat of particular species. Those which are parasitic on algae, other fungi, or invertebrates can be found by collecting material of the host and examining it under the microscope. This is a rather slow process and isolating such species is often difficult.

The most common method of isolating zoosporic aquatic fungi is through the use of suitable substrates called "baits." Among the more common baits are hemp seed, pollen (*Liquidambar styraciflua* and *Pinus* spp. work well), cellophane, or dialyzing membrane (cellulose), human hair that has been defatted in ether, snake skin, or thin slices of nails or cow horn (keratin), and shrimp exoskeleton (chitin). Dead flies or pieces of boiled paspalum grass also may be used. These baits may be used with either water or plant debris (leaves, twigs, fruits) from lakes and ponds, or with soils.

To set up plates for isolation place a small amount of vegetable debris or about 2 tbls. of soil in a petri plate, then add sterile distilled water to a depth of about 5 mm. Do not add too much debris or soil as that will cause the water to stagnate and you will get little out of it. If you are starting with pond or lake water, simply pour the water into the petri plate. Now add *small* quantities of the baits to the surface of the water; you can use several baits per plate but try not to mix them as this makes observation more difficult. It is better to use more plates with less bait per plate. Sprinkle a small amount of pollen on the water surface. Other baits, such as shrimp exoskeleton or cellophane, should be cut into squares small enough to fit under a cover glass when they are examined under the microscope. Place a few of these squares in the water; they can be sterilized by dipping quickly in boiling water if desired. Hemp seed should be cut in half with a scalpel or razor blade before placing in the water. The seeds cut better if they are dipped briefly in boiling water to soften the seed coat; this will also sterilize them.

Set the baited plates aside for 3-4 days, then examine some of the baits under the microscope. Pollen can be picked up from the water surface by touching it with a cover glass or the end of a needle handle. Then dip the needle handle in a drop of water on a slide and the pollen will come off. If a cover glass is used it can be placed directly on a slide. Squares of cellophane or similar materials should be mounted in a drop of water on a slide. Examine the baits carefully; remember that the simplest species consist of only a single, hyaline, spherical cell.

To obtain unifungal cultures of the organisms you find, place some of the bait bearing them in fresh plates of sterile distilled water and add fresh bait. Within a few days the new bait should contain the fungus. You may have to repeat this several times to obtain pure cultures; bacterial contamination is often a problem.

Filamentous zoosporic fungi usually develop on hemp seed and flies and they can be recognized by their larger size and abundant hyphae growing out of the bait. The entire bait, or portions of mycelium, may be transferred to a sterile water flask containing fresh bait, and pure cultures may be obtained.

There are numerous specialized techniques for isolating particular fungi. Blastocladiaceous fungi, for example, usually occur on submerged twigs and rosaceous fruits. To isolate these fungi, place small pieces of twig (especially ash, birch, apple or pear) or small rosaceous fruits (crabapple, rose, small apples) in wire baskets and submerge in a suitable lake or pond. After several weeks, remove and take to the laboratory. Wash off the slime surrounding the fruits (this should be examined for fungi, also) and look for pustules of fungi on the surface of the fruit. These will usually contain blastocladiaceous species. Water can also be placed in large jars in the laboratory, but this is usually not as satisfactory.

Phytophthora may be isolated using the apple technique. With a cork borer, bore a hole approximately 1 cm. in diameter nearly through a healthy apple. Pack the hole tightly with soil to within 1 cm. of the surface of the apple, then fill the rest of the hole with distilled water to moisten the soil. Seal the hole with scotch tape and incubate at room temperature. If *Phytophthora* species are present, they will invade the apple tissue, causing a firm dry rot that is brown in color. Five to seven days after inoculation, remove small pieces of rotted apple (not in contact with the soil) and plate on cornmeal agar. Other soil fungi may also invade the apple tissue, but they usually develop more slowly, after 8 to 10 days and cause a soft rot. The apple cultivar 'Golden Delicious' presumably works best, but other cultivars can be used.

Many species of aquatic fungi can be grown on agar, especially on yeast extract — starch medium (YPSS), which was devised for this purpose. Pure cultures may be maintained on the substrate in sterile water in small flasks; these usually need to be transferred fairly frequently.

Isolation of Zygomycetes

The mucoraceous Zygomycetes are among the most common of fungi. They occur on all types of substrates, including dung and decaying fruits, where they produce a profuse, rapidly growing mycelium. Instead of zoospores they form sporangiospores that are usually borne on erect sporangiosphores. Most can be readily isolated and grown in pure culture by transferring sporangia or spores to a suitable medium with a sterile needle. Allow the needle to cool so it will not incinerate the sporangia. Mycelium may also be used; aerial mycelium not in direct contact with the substrate is less likely to be contaminated with bacteria. Malt extract and V-8 juice agars are good media for culturing most mucoraceous fungi.

Isolation of Ascomycetes and Deuteromycetes

Ascomycetes and deuteromycetes are among the most ubiquitous of fungi, with the latter usually being more abundant. Some sources of these fungi and isolation methods are given below.

Natural substrates — Ascomycetes and deuteromycetes are both common in natural substrates and may be readily isolated from them. Pieces of wood, decaying leaves, dung, or similar materials can serve as good sources for these fungi. These materials can be placed either on agar or in a moist chamber and incubated at room temperature for several days. As various fungi appear they can be removed and grown in pure culture, although the profusion of fungi often makes it difficult to separate out different species. Sometimes conidia can be picked off by touching them with a moist sterile needle, or a needle dipped in agar. Perithecial species can often be isolated by inverting an agar plate over a moist clump of ascocarps. As ascospores are discharged they will stick to the agar. After a short period the agar plate should be removed and covered with a sterile cover.

Cellulose decomposing fungi can be isolated by laying strips of sterile filter paper over the sample. The fungi can then be isolated after they have grown out on the paper.

Yeasts can often be isolated from the surface of ripe fruits, which have been placed in a damp chamber for a few days. Direct isolation can be made from colonies that develop on the fruit. Another useful technique is to wash the surface of the fruit in a few ml of sterile water and plate out the water, diluting it if necessary.

Internal fungi – Most plant tissues, such as seeds, leaves, and stems, contain an internal fungus flora, even though they show no symptoms. To isolate internal fungi, the tissue should first be surface disinfected to remove external spores and mycelium. To do this, cut the sample into small (ca 1 cm^2) pieces and immerse for 2 min in a solution containing 10 ml commercial bleach (sodium hypochlorite), 10 ml ethyl alcohol, and 80 ml water. The alcohol acts to reduce surface tension and give better coverage of rough surfaces. Some people prefer to dip the material briefly in 95% alcohol, then follow this with immersion in the bleach. After immersion, place the pieces on the surface of suitable agar media, such as Martin's rose bengal — streptomycin medium. For materials with rough or pubescent surfaces the amount of bleach and alcohol are sometimes increased to 20 ml each, and 60 ml water.

The above procedure can also be used to isolate pathogenic fungi from diseased tissues. Such fungi are usually isolated most readily from tissues removed from the margins of lesions or galls.

Soils – Soils are one of the best sources of fungi, as they contain large numbers of them. Fungi occur in all soils, but those with high levels of organic matter have the greatest diversity of species. Below are some methods of isolating soil fungi. The isolation of ascomycetes can often be enhanced by treating the soil sample with alcohol or heat before plating it. To treat with alcohol, pour 5 ml of 60% ethanol over 5 gm soil and leave for 8-10 minutes before diluting or plating. Moistened soil can be heat treated at 60-70°C for 15-60 minutes.

Direct Inoculation

A crumb of soil is transferred with sterile forceps to the center of a petri dish containing an agar medium. Mycelia present in the crumb will grow out onto the agar and they can then be transferred to agar slants.

Soil Plate

To prepare a soil plate transfer a small amount of soil to a sterile petri plate with sterile forceps or spatula. Crush any soil aggregates. Add to the petri dish 8-10 ml of melted agar medium and disperse the soil particles throughout the medium. The agar should be cooled to just above the solidification point (ca 45°C) to avoid killing organisms in the soil. With heavy soils it may be necessary to mix the soil particles with a few drops of sterile water in the plate before addition of the agar, to facilitate their dispersal in the agar.

Soil Dilution Plate

The soil dilution plate is widely used both for determining relative numbers of organisms in the soil and for isolating pure cultures. The procedure given below is a general one and can be varied to suit individual needs.

1. Soil to be diluted is sifted through a 9 — mesh sieve. Three aliquots (5-10 gm each) of soil from the sample to be diluted are placed in previously weighed containers, weighed, and dried overnight in an oven (105-110°C). The aliquots are reweighed and the moisture content calculated.

2. A 25 gm sample of soil (on a dry-soil basis) is placed in a graduated cylinder. Water is added to the soil until a total of 250 ml is reached. The suspension is stirred and poured into a 1 liter Erlenmeyer flask. The flask is shaken on a mechanical shaker for 30 min. A Burrell "wrist-action" shaker (position 2-3) is recommended.

3. Ten ml of this suspension are immediately drawn (while in motion) into a sterile 10 ml pipette and transferred to a 90 ml sterile water blank. Eight-ounce, graduated, narrow mouth, screw cap medicine bottles are preferred for this.

4. Ten ml samples are then transferred immediately through successive 90 ml sterile water blanks until the desired final dilution is reached. Each suspension is shaken by hand for a few seconds, and is in motion while being drawn into the pipette. The sample should not be allowed to remain in any dilution for more than 10 min.

5. Use of the preceding method for making soil dilutions will yield dilutions of soil in water of 1 in 10, 1 in 100, 1 in 1,000, 1 in 10,000, 1 in 100,000, 1 in 1,000,000, 1 in 10,000,000, etc. Intermediate dilutions may be obtained by varying the volume of the water blank and the amount transferred. For example, if a dilution of 1 in 20,000 is desired, transfer 5 ml of the 1 in 1000 dilution to a 95 ml water blank.

6.(a) One ml of the desired dilution is transferred aseptically into each of several petri dishes, and 12-15 ml of an appropriate agar medium, cooled to just above the solidifying temperature, are added to each dish. The dishes are rotated by hand in a broad, swirling motion so that the diluted soil is dispersed in the agar medium. After incubation at 24-30°C, usually 5-7 days for fungi and 6-14 days for bacteria and actinomycetes, the resulting colonies are counted. For counting purposes, dishes containing fungal or bacterial spreaders or large clear zones of antagonism should be discarded. The average number of colonies per dish is multiplied by the dilution factor to obtain the number per gram in the original soil sample. Single colonies can be transferred to agar tubes for further study.

6.(b) An alternative method for isolating fungi; 1 ml of the final dilution is distributed over a solidified agar medium which has been poured 2-3 days previously.

A variety of media can be used in soil dilution plates; Martin's rose bengal-streptomycin medium is especially good.

Determining Soil pH

It is often desirable in ecological studies to determine the pH of the soil. To do this, add 20 ml distilled water to 20 gm soil (or 1:1 ratio by wt.). Stir and let stand for 30 min. Stir again and take pH reading while suspension is in motion. pH determinations should be made as soon as feasible after collecting the sample. If samples must be stored, keep moist and store at low temperature.

Isolation of Basidiomycetes

Most basidiomycetes do not form fruiting bodies in culture, although many will produce a mycelium. Consequently it is often difficult to identify basidiomycetes solely on the basis of cultures.

Some smut fungi can be cultured by placing spores on agar and allowing them to germinate. The resulting mycelium can then be transferred to fresh agar. If spores are removed from sori just before they open, the chances of contamination are reduced.

Mushrooms and bracket fungi can often be isolated by removing pieces of tissue from the fruit body and plating them on agar. Conditions should be as sterile as possible to reduce the chance of contamination. Both cap and stipe tissues can be used; which gives the best results varies with the species. Generally, tissue from actively growing areas of young sporocarps will give best results.

Wood rotting basidiomycetes can be isolated by taking small pieces of wood from the edge of the decayed area and plating them on agar. Malt agar is frequently used to grow both agarics and wood-rotting species.

REFERENCES

Booth, C. (Ed.). 1971. *Methods in Microbiology.* Vol 4. Academic Press, N.Y.

CMI. 1960. *Herb. I.M.I. Handbook.* Commonwealth Mycol. Inst., Kew, England.

Dhingra, O.D., and J.B. Sinclair. 1985. *Basic Plant Pathology Methods.* CRC Press, Inc., Boca Raton.

Fuller, M.S., and A. Jaworski. (Eds.). 1986. *Zoosporic Fungi in Teaching and Research.* Palfrey Contrib. in Botany, No. 3, Dept. of Botany, Univ. Georgia, Athens.

Hawksworth, D.L. 1974. *Mycologist's Handbook.* Commonwealth Mycol. Inst., Kew, England.

Johnson, L.F., and E.A. Curl. 1972. *Methods for Research on the Ecology of Soil-Borne Pathogens.* Burgess Publ. Co., Minneapolis.

Johnston, A., and C. Booth. (Eds.) 1983. *Plant Pathologist's Pocketbook.* 2nd ed. Commonwealth Mycol. Inst., Kew, England.

Stevens, R.B. 1974. *Mycology Guidebook.* Univ. Washington Press, Seattle.

Tuite, J. 1969. *Plant Pathological Methods. Fungi and Bacteria.* Burgess Publ. Co., Minneapolis.

Composition of Media and Solutions Used in Introductory Mycology

Since relatively few of the thousands of species of fungi will grow and sporulate readily in pure culture, hundreds of media have been devised to induce them to do so. Only a few of these media are used in this course; the formulae for these are given below.

Culture media fall into broad categories: natural and synthetic. Natural media include substrates such as herbaceous or woody stems, seeds, leaves, corn meal, oat meal, and wheat germ. These substances may either be incorporated into an agar base, placed on top of the agar, or used alone, usually in a test tube or petri plate containing a small amount of water. Natural media are especially useful for culturing fungi whose nutrient requirements are unknown or are unimportant. They are usually easy to prepare but they have the disadvantage that their composition is largely unknown.

Synthetic media are desirable whenever the composition of the medium must be known, as in physiological studies. Their advantage is that the ingredients can be duplicated. Czapek-Dox medium would fall in this category.

All media used in culturing fungi must be sterilized before use. Dry materials, such as stems, can be sterilized with propylene oxide gas. This is done as follows: place the material to be sterilized in a test tube or petri dish, then put this in a dessicator or other sealable glass container of suitable size. Make sure that air can penetrate the tube or dish containing the material to be sterilized. Add propylene oxide to the container at the rate of 1 ml per liter capacity of the container, seal, and leave at least 24 hours. After removal from the propylene oxide allow the material to air for another day before using, to avoid any residual action. Propylene oxide is toxic, very volatile, highly flammable, and must be handled carefully. It should be refrigerated during storage. If possible, carry out sterilization in a hood. Gaseous sterilization can also be used for agar plates.

Liquids can be sterilized with seitz or millipore filters. This is necessary when working with heat sensitive materials, such as antibiotics.

Steam sterilization by autoclaving is the customary method of sterilizing most culture media, but it cannot be used with heat labile compounds. Autoclaving may also produce undesirable chemical changes in natural substrates, rendering them ineffective as culture media. Generally, however, it can be used for sterilizing all media and substrates used for growing fungi. Such materials should be autoclaved for 15-20 minutes at 17 pounds pressure. It is not necessary to dissolve the ingredients before sterilization, unless agar slants are to be made, as they will dissolve during autoclaving. Less condensation will form in the petri dishes if the agar is cooled some before pouring the plates.

When autoclaving media do not fill flasks too full or they will boil over. A convenient method to avoid this is to make one-half liter of medium in a one liter flask. To do this simply add one-half of the amount of each ingredient given in the formulae to a one liter flask. One-half liter of medium will make approximately 20 petri plates, depending upon how thick you pour them.

For some fungi that do not compete well on highly nutritious media, half-strength medium can be made. This is done primarily with natural media, such as corn meal or rabbit food agars. Half strength plates are marked with the medium abbreviation plus /2, as CMA/2 or RFA/2.

CULTURE MEDIA

Corn Meal Agar (CMA)

Corn meal	20 gm
Peptone	20 gm
Dextrose	20 gm
Agar	15 gm
Distilled water	1,000 ml

Add the corn meal to the water and simmer for 30 min. to one hr. Filter through several layers of cheesecloth. Add the other ingredients to the filtrate, bring up to volume with distilled water, then autoclave.

Corn meal agar is frequently used alone, i.e., without the peptone and dextrose. In this form it has relatively low nutritive value and is very useful for inducing sporulation in many fungi. It is especially useful in obtaining ascocarps in members of the Gymnoascaceae. The amount of corn meal can be varied, as can any additives.

Czapek's (Czapek-Dox) Agar (CZA)

$NaNO_3$	3.0 gm
K_2HPO_2	1.0 gm
$MgSO_4 \cdot 7H_2O$	0.5 gm
KCl	0.5 gm
$FeSO_4 \cdot 7H_2O$	0.01 gm
Sucrose	30.0 gm
Agar	15.0 gm
Distilled water	1,000.0 ml

This is a synthetic medium commonly used in the growth of soil fungi. It is the primary medium upon which the manuals of *Aspergillus* and *Penicillium* are based. In general the high osmotic concentration of this medium reduces the amount of mycelial growth obtained with most fungi. Many species look quite different on Czapek's Agar when compared to other media. The fact that it is a synthetic medium assures ready duplication; this is especially desirable for physiological studies.

Glucose-Peptone Agar (GPA)

Glucose	5 gm
Peptone	1 gm
Agar	20 gm
Distilled water	1,000 ml

This medium is useful for the isolation of cellular slime molds from soil. The ingredients can be varied as desired.

Hemp Seed Agar (HSA)

Hemp seed extract	30 ml
Agar	15 gm
Distilled water	750 ml

To prepare hemp seed extract use 500 gm hemp seed and one liter of water. Place these in a blendor, small amounts at a time, and switch blendor on-off 10 times per batch. Autoclave mixture for 30 minutes, then strain through cheesecloth. Add water to filtrate to bring to one liter volume. While stirring, pipette 30 ml portions into test tubes. Use one tube of extract for each batch of agar.

Hemp seed agar is used mainly for growing pythiaceous fungi.

Hay Infusion Agar (HIA)

Hay	2.5 gm
$K_2HPO_4 \cdot 3H_2O$	2.0 gm
Agar	20.0 gm
Distilled water	1,000.0 ml

Steep hay in water for 15-30 minutes, then filter through cheesecloth. Add potassium hydrogen phosphate and agar and autoclave.

This is a very weak nutrient medium that is especially good for isolating protostelids.

Lactose-Peptone Agar (LPA)

Lactose	1 gm
Peptone	1 gm
Agar	20 gm
Distilled water	1,000 ml

This is another low nutrient medium for isolating cellular slime molds

Lactose-Yeast Extract Agar (LYA)

Lactose	1.0 gm
Yeast extract	0.5 gm
Agar	20.0 gm
Distilled water	1,000.0 ml

This low nutrient medium is especially good for isolating protostelids.

Malt Extract Agar (MEA)

Malt extract	20 gm
Peptone	1 gm
Dextrose	20 gm
Agar	20 gm
Distilled water	1,000 ml

This is another good general medium. It was originally devised by Blakeslee for the mucors and is one of the media used in the identification of *Penicillium* and *Aspergillus*. The peptone and dextrose are often varied in proportion or omitted entirely.

Wood rotting fungi are often grown on a modification of this formula, as follows:

Malt extract	25 gm
Agar	15 gm
Distilled water	1,000 ml

Malt-Salt Agar (MSA)

Sodium chloride	68 gm
Malt extract	10 gm
Agar	20 gm
Distilled water	1,000 ml

This medium is used for growing species requiring a high osmotic concentration. The amount of salt can be varied as desired.

Harrold's Agar (M4OY)

Sucrose	400 gm
Malt extract	20 gm
Yeast extract	5 gm
Agar	20 gm
Distilled water	1,000 ml

This medium is used for isolating and growing osmophilic fungi. The sugar content can be varied as needed. The sugar will dissolve better if the flask is allowed to sit for an hour or so before autoclaving.

Potato Dextrose Agar (PDA)

Peeled, diced potatoes	200 gm
Dextrose	20 gm
Agar	15 gm
Distilled water	1,000 ml

Prepare the potatoes as for corn meal. The amount of potatoes used can be varied. Some workers use instant mashed potatoes instead of fresh potatoes.

Potato dextrose agar supports the growth of many fungi and has long been a favorite medium among plant pathologists for culturing plant pathogenic fungi.

Potato Sucrose Agar (PSA)

Prepare as for PDA, but substitute sucrose for the dextrose.

Rose Bengal Medium (RBM-2)

KH_2PO_4	0.5 gm
K_2HPO_4	0.5 gm
$MgSO_4 \cdot 7H_2O$	0.5 gm
Peptone	0.5 gm
Dextrose	10.0 gm
Yeast extract	0.5 gm
Rose bengal	0.05 gm
Streptomycin	0.03 gm
Agar	17.0 gm
Distilled water	1,000.0 ml

Make up the medium without the streptomycin and autoclave. Since streptomycin is heat labile, it must be added just before the agar solidifies (ca 45°C). The easiest way to do this is to make up a stock solution of streptomycin sulfate and sterilize with a seitz or millipore filter. Then pipette the proper amount of solution into the flask just before pouring.

This is an excellent medium for isolating fungi from soils and plant parts. The streptomycin retards bacterial growth and the rose bengal slows down the growth of some of the faster growing fungi.

There are several modifications of this medium; the formula above is taken from Tsao (1964).

Rabbit Food Agar (RFA)

Rabbit food	50 gm
Agar	20 gm
Distilled water	1,000 ml

Steep the rabbit food in 600 ml water for 30 minutes, then filter through cheesecloth. Bring to 1 liter volume, add agar, and autoclave.

This is a good medium for growing certain imperfect fungi, especially wood inhabiting forms. When diluted about 1:4 with water, it is a good substitute for hay infusion for isolating acrasiomycetes.

Tannic (Gallic) Acid Agar (TA) (GA)

Tannic (or gallic) acid	5 gm
Malt extract	15 gm
Agar	20 gm
Distilled water	1,000 ml

Autoclave malt and agar in 850 ml water. Sterilize other 150 ml water in separate flask. When done, add tannic acid to pure water while still hot, and dissolve it. Then add to agar-malt flask, mix, and pour plates.

This medium is used to determine the oxidase production of wood rotting fungi.

Vegetable Juice Agar (V-8)

V-8 Juice	180 ml
Calcium carbonate	2 gm
Agar	20 gm
Distilled water, to	1,000 ml

V-8 juice agar is an excellent general purpose agar for routine use. Most ascomycetes sporulate well on it, as do the mucors and many Fungi Imperfecti. The medium is fairly acid (ca pH 5.5). The percentage V-8 juice is not critical; 180 ml (18%) is routinely used because this is the size of a small can of V-8 juice. Other juices, such as tomato or carrot, can be substituted for V-8 juice.

Yeast-Starch Agar (YPSS)

Yeast extract	4.0 gm
Soluble starch	15.0 gm
K_2HPO_4	1.0 gm
$MgSO_4 \cdot 7H_2O$	0.5 gm
Agar	20.0 gm
Distilled water	1,000.0 ml

This medium was developed by Emerson for *Allomyces*, but it is also good for other fungi, especially zoosporic fungi.

Water Agar (WA)

Agar	20 gm
Distilled water	1,000 ml

Water agar is useful where sparce mycelial growth is desirable, as in isolating fungi from soil or plant parts. It can also be used to induce sporulation in some fungi.

In the media described above you will notice that the amount of agar varies. This is a matter of personal preference and is not critical. It is customary to use between 15 gm (1.5%) and 20 gm (2.0%) agar per liter. The more agar used the harder the surface of the poured petri plate will be. Larger or smaller amounts are also sometimes used for special purposes.

Dehydrated Media

Several companies produce dehydrated agar media in concentrated powder form. The powder is added to water and autoclaved to prepare the medium. Our experience is that dehydrated media do not give exactly the same reaction as freshly prepared media involving natural materials such as potatoes. They do permit one to stock a variety of media without buying the individual ingredients, but at a somewhat higher cost. Dehydrated media are especially useful when only small amounts are needed or use is occasional.

Companies manufacturing dehydrated media are as follows: Baltimore Biological Laboratory, Inc., 2201 Aisquith St., Baltimore, Md., Ben Venue Laboratories, Inc., Bedford, Ohio. Media in tablet form. Consolidated Laboratories, Inc., Chicago Heights, Ill. Oxoid and Colab brand media. Difco Laboratories, Inc., P.O. Box 1058B, Detroit, Mich. 48232. This is the oldest company with the largest selection of media. Many publications specify "Difco" products. Fisher Scientific, 690 Miami Circle, N.E., Atlanta, Ga. Fisher features the "Gram-pak" for making small quantities of medium.

SOLUTIONS

Amman's Lactophenol Solution

Phenol	20 ml
Lactic acid	20 ml
Glycerine	40 ml
Distilled water	20 ml

This solution can be used to make semi-permanent slide preparations from living material. When you have a water mount you wish to preserve, place one or two drops of lactophenol solution along one edge of the cover glass. With a small piece of paper towel or bibulous paper, draw off the water from the opposite edge of the cover glass, so that the lactophenol will replace the water. Carefully wipe away excess lactophenol, then ring the cover glass several times with nail polish to seal it. Properly prepared slides will remain in good condition for years.

Lactophenol solution is sometimes prepared with a stain in it, such as cotton blue or phloxine.

Melzer's Reagent

Iodine crystals	0.5 gm
Potassium iodide	1.5 gm
Chloral hydrate	20.0 gm
Distilled water	20.0 ml

This solution is used on fungal tissues, especially ascus tips and basidiospores. A positive reaction gives a dark blue (amyloid) stain. Intermediate reactions of yellowish- or reddish-brown (dextrinoid) are also obtained with some species. The natural color of the iodine must be taken into account when interpreting these reactions.

Potassium Hydroxide

Since the tissues of fleshy fungi such as mushrooms and discomycetes shrink during drying, they need to be swollen to normal size for study. This is usually done by mounting thin sections of the fruiting body in a 2-3% aqueous solution of potassium hydroxide (KOH). After a few minutes the tissues will rehydrate and be ready for examination. If desired, a small amount of a dye, such as phloxine, can be added to the KOH to stain the tissues.

Kohn and Korf (1975) recommend that tissues to be treated with Melzer's Reagent be routinely rehydrated in KOH, as this procedure gives more consistent results than rehydration with water.

STAINS

There are many stains that can be used with fungi; we use two: cotton blue and phloxine. Both are made up as 0.5% aqueous solutions and a drop or two placed alongside the cover glass is sufficient for most materials.

REFERENCES

Consolidated Laboratories. 1961. *Colab Products for the clinical laboratory. Oxoid culture media and laboratory preparations.* Consolidated Labs, Inc., Chicago Hgts., Ill.

Cote, R. (ed.). 1984. ATCC Media Handbook. American Type Culture Collection, Rockville.

Difco Laboratories. 1953. *Difco manual of dehydrated culture media and reagents for microbiological and clinical laboratory procedures.* 9th ed. Difco Laboratories, Detroit, Mich.

Difco Laboratories. 1966. *Difco supplementary literature.* Difco Labs, Detroit, Mich.

Fisher Scientific. 1963. *Fisher bacteriological culture media.* Fisher Scientific, 690 Miami Circle, N.E., Atlanta, Ga.

Hansen, H.N., and W.C. Snyder. 1947. Gaseous sterilization of biological material for use as culture media. *Phytopathology* 37: 369-371.

Kohn, L.M., and R.P. Korf. 1975. Variation in ascomycete iodine reactions: KOH pretreatment explored. *Mycotaxon* 3: 165-172.

Stevens, R.B. 1974. *Mycology Guidebook.* University Washington Press, Seattle.

Tsao, P.H. 1964. Effect of certain fungal isolation agar media on *Thielaviopsis basicola* and on its recovery in soil dilution plates. *Phytopathology* 54: 548-555.

Tuite, J. 1978. Natural and synthetic media: fungi. p. 427-504. In M. Rechcigl, Jr. (ed.), *CRC Handbook Series in Nutrition and Food, Section G: Diets, Culture Media, Food Supplements.* Vol. III. Culture Media for Microorganisms and Plants. CRC Press, Inc., Cleveland.

Mycology Culture Collection

An important part of mycology is knowing where to look for the various kinds of fungi and how to isolate them in pure culture once you have found them. Not all fungi will grow in culture, but many will. Some that will grow do not readily produce spores or fruiting structures, so their identification is difficult. Others can be grown only through the use of specialized media and techniques. In order to give you some experience in the isolation of fungi from natural substrates, each student will be required to make a culture collection of fungi, to be turned in at the end of the term.

The culture collection should be your own work; please do not trade cultures or use class materials from this or any other course. Each culture must contain only one fungus and must be free of bacteria, except for members of the Myxomycota.

Each of the isolates must represent a different genus and should include the following:

Undergraduate	*Graduate*
2 Aquatic fungi or Zygomycete	2 Aquatic fungi or Zygomycete
1 Ascomycete	2 Ascomycetes (must have asci)
2 Fungi Imperfecti	3 Fungi Imperfecti
1 Your choice	2 Your choice
—	—
6	9

There are many isolation techniques that can be used; some of these are outlined on preceding pages. You should try various methods of isolation as they will influence what fungi you find. If you are interested in some particular area you can emphasize this in your techniques and selection of materials. Any special media required will be provided, but you must allow time for preparation.

All cultures must be turned in by the last week of class; you may turn them in earlier if you wish. All isolates should be on agar slants, except for the zoosporic fungi. These may be turned in as water cultures in small flasks. Be certain that all cultures are clearly labelled with your name and the name or number of the isolate. Along with the cultures you must prepare a report to accompany them. The report must contain the following information for each isolate:

(a) Identification to class, order, family, and genus. You should also indicate the species, if feasible.
(b) References used in identifying fungus.
(c) Isolation method. Be precise.
(d) Nature and source of substrate. Be specific.
(e) Accurately labelled drawings of all reproductive structures.
(f) Semi-permanent slide showing reproductive structures of the fungus.

Slides may be made semi-permanent by replacing the water with lactophenol and sealing the cover glass with at least 2 applications of nail polish.

Slides, cultures, and reports will be returned to you provided they are picked up soon after the term ends.

The culture collection should be started as early in the term as possible, since some fungi, especially Ascomycetes, are difficult to isolate. Becoming familiar with isolation techniques and identifying the fungi you isolate will be slow at first. Time will also be needed to subculture your isolates and grow them in pure culture, so it is to your advantage to begin early.

GENERAL REFERENCES FOR THE IDENTIFICATION OF FUNGI

Ainsworth, G.C., F.K. Sparrow, and A.S. Sussman. (eds.). 1973. *The Fungi — An Advanced Treatise. Vol. IVA. A Taxonomic Review with Keys: Ascomycetes and Fungi Imperfecti.* Academic Press, New York.

Ainsworth, G.C., F.K. Sparrow, and A.S. Sussman. (eds.). 1973. *The Fungi — An Advanced Treatise. Vol IVB. A Taxonomic Review with Keys: Basidiomycetes and Lower Fungi.* Academic Press, New York.

Arx, J.A. von. 1981. *The Genera of Fungi Sporulating in Pure Culture.* 3rd. ed. J. Cramer, Vaduz.

Barnett, H.L., and B.B. Hunter. 1972. *Illustrated Genera of Imperfect Fungi.* 3rd ed. Burgess Publ. Co., Minneapolis.

Barron, G.L. 1968. *The Genera of Hyphomycetes from Soil.* Williams and Wilkins Co., Baltimore.

Breitenbach, J., and F. Kränzlin. 1981. *Fungi of Switzerland.* Vol 1. *Ascomycetes.* Verlag Mykologia Lucerne, Lucerne.

Dennis, R.W.G. 1977. *British Ascomycetes.* 3rd. ed. J. Cramer, Lehre.

Domsch, K.H., W. Gams, and T.-H. Anderson. 1983. *Compendium of Soil Fungi.* Vols. 1 & 2. Academic Press, New York.

Ellis, M.B., and J.P. Ellis. 1985. *Microfungi on Land Plants.* Macmillan Publ. Co., New York.

Guzmán, Huerta, G. 1977. *Identificacion de los Hongos Comestibles, Venenosos, Alucinantes y Destructores de la Madera.* Editorial Limusa, México, D.F.

Hale, M.E. 1969. *How to Know the Lichens.* Wm. C. Brown Publ., Dubuque.

Hawksworth, D.L., B.C. Sutton, and G.C. Ainsworth. 1983. *Ainsworth & Bisby's Dictionary of the Fungi.* 7th ed. Commonwealth Mycol. Inst., Kew.

Hesler, L.R. 1960. Mushrooms of the Great Smokies. Univ. Tennessee Press, Knoxville.
Karling, J.S. 1977. *Chytridiomycetarum Iconographia*. J. Cramer, Vaduz.
Miller, O.K., Jr. 1972. *Mushrooms of North America*. E.P. Dutton & Co., Inc., New York.
Richardson, M., and R. Watling. 1968. *Keys to Fungi on Dung*. Bull. British Mycol. Soc. 2: 18-43.
Smith, H.V., and A.H. Smith. 1973. *How to Know the Non-gilled Fleshy Fungi*. Wm. C. Brown Co., Dubuque.
Sparrow, F.K., Jr. 1960. *Aquatic Phycomycetes*. 2nd ed. Univ. Michigan Press, Ann Arbor.

GENERAL REFERENCES FOR INTRODUCTORY MYCOLOGY

Alexopoulos, C.J., and C.W. Mims. 1979. *Introductory Mycology*. 3rd. ed. John Wiley & Sons, New York.
Beckett, A., I.B. Heath, and D.J. McLaughlin. 1974. *An Atlas of Fungal Ultrastructure*. Longman, Ltd., London.
Burnett, J.A. 1976. *Fundamentals of Mycology*. 2nd ed. St. Martin's Press, N.Y.
Deacon, J.W. 1984. *Introduction to Modern Mycology*. 2nd ed. Blackwell Sci. Publ., London.
Dube, H.C. 1983. *An Introduction to Fungi*. Vikas Publ. House, New Delhi.
Kendrick, B. 1985. *The Fifth Kingdom*. Mycologue Publ., Waterloo.
Koch, W.J. 1966. *Fungi in the Laboratory*. Book Exchange, Chapel Hill, N.C.
Moore-Landecker, E. 1982. Fundamentals of the Fungi. 2nd ed. Prentice-Hall, Inc., Englewood Cliffs, N.J.
Ross, I.K. 1979. *Biology of the Fungi*. McGraw-Hill Book Co., New York.
Talbot, P.H.B. 1971. *Principles of Fungal Taxonomy*. Macmillan Press, London.
Tsuneda, A. 1983. *Fungal Morphology and Ecology*. Tottori Mycol. Inst., Tottori.
Webster, J. 1980. *Introduction to Fungi*. 2nd ed. Cambridge University Press, London.
Wolf, F.A., and F.T. Wolf. 1947. *The Fungi*. Vols. 1 & 2. John Wiley & Sons, N.Y.

Kingdom Fungi

The organisms included under the general category of "fungi" are so diverse that it is difficult to characterize them precisely. All are heterotrophic and absorptive. The thallus varies from an amoeboid myxamoeba or plasmodium lacking cell walls, to a unicellular or filamentous thallus with a rigid cell wall. In species with a filamentous (mycelial) thallus, the thallus may be septate or non-septate, but those with septa are functionally coenocytic, since the septa are perforate. Thalli may be in or on the host or substrate. The thallus wall is typically chitinous, but occasionally it is cellulosic. Only rarely do chitin and cellulose occur together. Fungi are usually non-motile, but some produce flagellate cells. The nucleus is eukaryotic. Thalli may be uni- or multi-nucleate, with homo- or hetero-karyotic, haploid, dikaryotic, or diploid cells. Both asexual and sexual reproduction occurs. Fungi occur worldwide, in all types of habitats, as saprobes, symbionts, or parasites.

Myxomycota

In the Myxomycota the assimilative phase consists of an amoeboid myxamoeba or plasmodium. These cells lack rigid cell walls, being bound only by the plasmalemma, and they may be uni- or multi-nucleate. They feed by ingestion of bacteria and spores. In reproduction individual cells may be converted into the fruiting structures, the cells may converge into a pseudoplasmodium from which the fruiting structure forms, or the plasmodium may be converted into the fruiting structures. There is cellular differentiation associated with development of the fructification. These organisms are commonly associated with decaying vegetable debris.

Acrasiomycetes

The assimilative phase of the Acrasiomycetes consists of free-living myxamoebae or minute plasmodia. In primitive forms the myxamoebae or plasmodia are converted into minute fruiting bodies (sporocarps) bearing one to few spores. In other species the myxamoebae aggregate into a pseudoplasmodium prior to formation of the fruiting body, or sorocarp.

Protosteliales

The vegetative stage consists of free-living myxamoebae or small plasmodia that feed on bacteria. Flagellate cells may or may not be present, depending upon the genus. Protostelids are especially common on dead plant structures such as flowers, seeds, pods and berries that are still attached to the plant. The fruiting bodies, or sporocarps, arise from individual protoplasts and consist of a slender stalk bearing a single spore (sometimes two spores in some species).

Protosteliaceae — Sporocarp formed from a single protoplast; flagellate cells lacking.

Protostelium — Examine sporocarps of *P. mycophaga* (Plate I) in culture under the dissecting scope.

Dictyosteliales

The Dictyosteliales, or cellular slime molds, have naked, free-living myxamoebae which feed on bacteria. Flagellate cells are lacking. During reproduction the myxamoebae aggregate to form a pseudoplasmodium and later a fruiting body, the sorocarp. The myxamoebae comprising the pseudoplasmodium do not fuse, but retain their individuality. Dictyostelids occur on all kinds of decaying vegetable matter but are especially common in surface soil and leaf litter of deciduous forests.

Dictyosteliaceae — The sorocarp stalk is cellular.

Dictyostelium – Members of this genus form usually unbranched sorocarps with a cellular stalk and globose spore-bearing sorus. Examine a culture of *D. discoideum* under the dissecting scope and observe stages in the life cycle. You should be able to see clumps of myxamoebae, pseudoplasmodia, and various stages in sorocarp formation.

Carefully scrape myxamoebae from the surface of the agar and mount in a drop of water. Gently cover with a cover glass to avoid crushing and observe the movement of the myxamoebae under the microscope.

Mount a pseudoplasmodium in water and watch the myxamoebae that comprise it as they separate into the individual cells.

Mount a sorocarp in water and observe the spores and the cellular stalk. (Plate I, Figs. 1-7).

Polysphondylium – Similar to *Dictyostelium*, but the sorocarp has regular whorls of branches bearing sori. Examine a culture under the dissecting microscope and observe the branched habit (Plate I).

REFERENCES

Barkley, David S. 1969. Adenosine — 3', 5'-phosphate: Identification as acrasin in a species of cellular slime mold. *Science* 165: 1133-1134.

Bonner, John T. 1963. How slime molds communicate. *Scientific American* 209: 84-93.

Bonner, John T. 1967. *The Cellular Slime Molds*. 2nd ed. Princeton Univ. Press, Princeton.

Cappuccinelli, P. (ed.). 1977. *Development and Differentiation in the Cellular Slime Molds*. Elsevier/No. Holland Biomedical Press, Amsterdam.

Cavender, J.C. 1973. Geographical distribution of the Acrasieae. *Mycologia* 65: 1044-1055.

Cavender, J.C. 1983. Cellular slime molds of the Rocky Mountains. *Mycologia* 75: 897-903.

Cavender, James C., and K.B. Raper. 1965. The Acrasiae in nature. I. Isolation. *Amer. J. Bot.* 52: 294-296.

Cavender, James C., and K.B. Raper. 1965. The Acrasiae in nature. II. Forest soil as a primary habitat. *Amer. J. Bot.* 52: 297-302.

Cavender, James C., and K.B. Raper. 1965. The Acrasiae in nature. III. Occurrence and distribution in forests of Eastern North America. *Amer. J. Bot.* 52: 302-308.

Cavender, James C., and K.B. Raper. 1968. The occurrence and distribution of Acrasieae in forests of subtropical and tropical America. *Amer. J. Bot.* 55: 504-513.

Lonert, A.C. 1965. A week-end with a cellular slime mold. *Turtox News* 43: 50-53.

Loomis, W.F. (ed.). 1982. *The Development of Dictyostelium discoideum*. Academic Press, New York.

Olive, L.S. 1967. The Protostelida — a new order of the Mycetozoa. *Mycologia* 59: 1-29.

Olive, L.S. 1974. *The Mycetozoans*. Academic Press, New York.

Raper, K.B. 1984. *The Dictyostelids*. Princeton Univ. Press, Princeton.

PLATE I. Left, Mature and developing sorocarps of *Dictyostelium discoideum*. × 20. Right, Mature branched sorocarps of *Polysphondylium violaceum*. × 20. Left lower center, Mature sorocarps of *Protostelium mycophaga*. × 120.

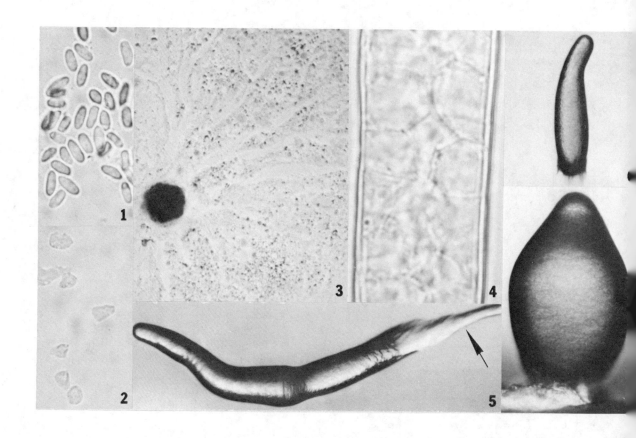

FIG. 1-7. *Dictyostelium discoideum*. FIG. 1. Mature spores. × 800. FIG. 2. Myxamoebae. × 400. FIG. 3. Movement of myxamoebae toward aggregation point (left). × 15. FIG. 4. Close-up of stalk showing cellular nature. × 1500. FIG. 5. Migrating pseudoplasmodium with slime trail (arrow). × 25. FIG. 6. Pseudoplasmodium at end of migration stage. × 20. FIG. 7. Early stage in sorocarp development. × 50.

Myxomycetes

The Myxomycetes are characterized by an amoeboid, free-living, saprobic plasmodium during the assimilative or vegetative stage. Several different kinds of plasmodia are recognized, but all are acellular and multinucleate, with diploid nuclei. Species with larger plasmodia have an extensive system of anastomosing veins in which the protoplasm streams. Plasmodia may be of many different colors and they vary in size from minute to quite large, depending upon the species. Under favorable conditions the plasmodium becomes converted into one or more fruiting bodies, or sporangia. The spores formed give rise to myxamoebae or swarm spores upon germination. Myxomycetes commonly inhabit moist, decaying vegetable debris, but some occur in dryer habitats, such as tree bark.

Ceratiomyxomycetidae

In the Ceratiomyxomycetidae the spores are borne singly and externally on short stalks on columnar or dendroid fruiting bodies. The subclass contains a single order, the **Ceratiomyxales**, with the single family **Ceratiomyxaceae**.

Ceratiomyxa fruticulosa — This species forms small clumps of pure white columnar branches on which the spores are borne. It is found on wet, decaying wood. Examine dried material under the dissecting microscope and observe the columnar nature of the fruiting body and the external spores. (Plate II).

Myxogastromycetidae

In this subclass the spores are borne inside various kinds of sporangia. The majority of Myxomycetes belong here. They are found in a variety of habitats.

Physarales

Spores dark in mass, ranging from black to violet-brown, dark purple-brown or deep red or purple. Sporophore development is subhypothallic. The peridium is typically calcareous; other parts of the sporophore may or may not contain lime. The assimilative stage is a phaneroplasmodium.

Physaraceae — Capillitium typically calcareous.

Physarum polycephalum — This species forms gyrose stalked sporangia. Observe a living plasmodium of this species under low power of the compound microscope and note the streaming of the protoplasm. With the dissecting microscope examine the stalked sporangia formed on filter paper (Plate II, Fig. 8).

Physarum cinereum — This species forms grayish sessile sporangia. It is common on lawns in summer. Observe the sporangia under the dissecting microscope, then make a water mount of spores and examine under the microscope.

Fuligo — Species in this genus form an aethalium and both capillitium and pseudocapillitium are present. Examine an aethalium under the dissecting microscope. Mount some spores of *Fuligo cinerea* in water and examine under the microscope (Plate II).

Trichiales

Spores pallid in mass, ranging from white or bright-colored to yellow- or smoky-brown; never purple-brown. Sporophore development is subhypothallic. The peridium is evanescent and no columella is present. Capillitium is usually abundant, thread-like, and sculptured.

Trichiaceae — Capillitial threads coarse, over 2μm in diameter, tubular, usually ornamented.

Arcyria — In this genus the capillitial threads are roughened. Examine a prepared slide of *Arcyria* sp. and note the capillitium and spores. Examine dried sporangia of available species under the dissecting microscope and note the differences (Plate II; Fig. 14).

Hemitrichia — The capillitial threads in *Hemitrichia* are ornamented with spiral bands. Examine prepared slides and note the capillitium and spores. Examine dried specimens of species with stalked sporangia and a plasmodiocarp (Plate II, Fig. 9-11).

Stemonitales

Spores violet-brown, lilac, ferruginous, or pallid by transmitted light. Sporophore development is epihypothallic. The peridium is usually evanescent at an early stage. True capillitium is typically present; lime, if present, is never on the capillitium. The assimilative stage is an aphanoplasmodium. There is a single family, the **Stemonitaceae.**

Comatricha — In this genus the sporangial peridium is usually evanescent but there is no surface net and there is no lime present. Examine dried sporangia under the dissecting microscope.

Diachea — In this genus there is conspicuous lime present in the sporangium. Examine dried sporangia under the dissecting microscope (Plate II).

Stemonitis — In *Stemonitis* the peridium is evanescent and a surface net is formed; no lime is present. Examine the stalked sporangia of this genus under the dissecting microscope. Examine prepared slides and note the spores, columella, and capillitium. Mount some spores in water and examine under the microscope (Plate II, Fig. 12).

Liceales

Spores pallid in mass, smoky- to yellow-brown, never purple-brown. Sporophore development is subhypothallic. The peridium is persistent. True capillitium is lacking; pseudocapillitium is present or not; if present, then of plate-like to tubular filaments.

Reticulariaceae — Sporophores sporangiate, pseudoaethalioid, or aethalioid; dictydine granules lacking.

Lycogala epidendrum — This species forms an aethalium with pseudocapillitium. Examine under the dissecting microscope, then mount some spores and pseudocapillitium in water and observe under the microscope (Plate II).

Cribrariaceae — As above, but dictydine granules present.

Dictydium — In this genus a net of stout longitudinal ribs is formed. Examine dried sporangia of *Dictydium cancellatum* under the dissecting microscope (Plate II).

REFERENCES

Aldrich, H.C., and J.W. Daniel. (eds.). 1982. *Cell Biology of Physarum and Didymium*. Vols. 1 & 2. Academic Press, New York.

Alexopoulos, C.J. 1963. The Myxomycetes. II. *Botanical Review* 29: 1-78.

Crowder, Wm. 1926, Marvels of Mycetozoa. *Nat. Geog. Mag.* 49: 421-443.

Dove, W.F., and H.P. Rusch. 1980. *Growth and Differentiation in Physarum polycephalum*. Princeton Univ. Press, Princeton.

Emoto, Y. 1977. *The Myxomycetes of Japan*. Sangyo Tosho Publ. Co., Ltd., Tokyo.

Farr, M.L. 1976. *Myxomycetes*. Flora Neotropica Monogr. 16. New York Botan. Gard.,

Farr, M.L. 1981. *How to Know the True Slime Molds*. Wm. C. Brown Co., Dubuque.

Gray, William D., and C.J. Alexopoulos. 1968. *Biology of the Myxomycetes*. Ronald Press Co., New York.

Howard, F.L. 1931. The life history of *Physarum polycephalum*. *Amer. J. Bot.* 18: 116-133.

Lakhanpal, T.N., and K.G. Mukerji. 1981. *Taxonomy of the Indian Myxomycetes*. J. Cramer, Vaduz.

Martin, G.W., and C.J. Alexopoulos. 1969. *The Myxomycetes*. University of Iowa Press, Iowa City.

Mims, C.W. 1971. An ultrastructural study of spore germination in the Myxomycete *Arcyra cinerea*. *Mycologia* 68: 586-601.

Sauer, H.W. 1982. *Developmental Biology of Physarum*. Cambridge Univ. Press,

Scheetz, R.W. 1972. The ultrastructure of *Ceratiomyxa fruticulosa*. *Mycologia* 69: 38-54.

PLATE II. Mature myxomycete sporangia. Top, l-r: *Physarum polycephalum*. × 10. *Dictydium cancellatum*. × 40. *Ceratiomyxa fruticulosa*. × 60. Middle, l-r: *Arcyria incarnata*. × 20. *Hemitrichia stipitata*. × 15. *Stemonitis nigrescens*. × 20. Bottom, l-r: *Diachea leucopodia*. × 40. *Fuligo cinerea* (aethalium). × 1. *Lycogala epidendrum* (aethalium). × 2.

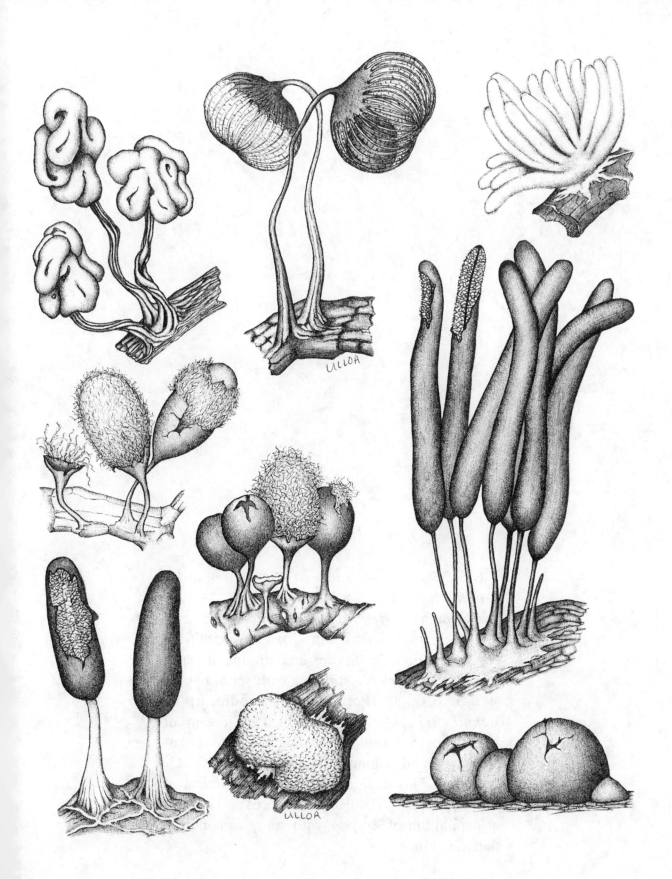

FIG. 8. Plasmodium of *Physarum polycephalum* on agar. × 0.65. FIG. 9. Mature sporangium of *Hemitrichia stipitata* with stalk (ST), remains of peridium (P), and capillitium (C). × 25. FIG. 10. Spores and strands of capillitium of *Hemitrichia stipitata* with spiral bands. × 625. FIG. 11. Portion of plasmodiocarp of *Hemitrichia serpula*. × 35. FIG. 12. Close-up of portion of *Stemonitis* sporangium with capillitium (C) and columella (CL). × 250. FIG. 13. Spores of *Plasmodiophora brassicae* in cells of cabbage root. × 625. FIG. 14. Spores and strands of capillitium of *Arcyria* with spiny ornamentations. × 1560.

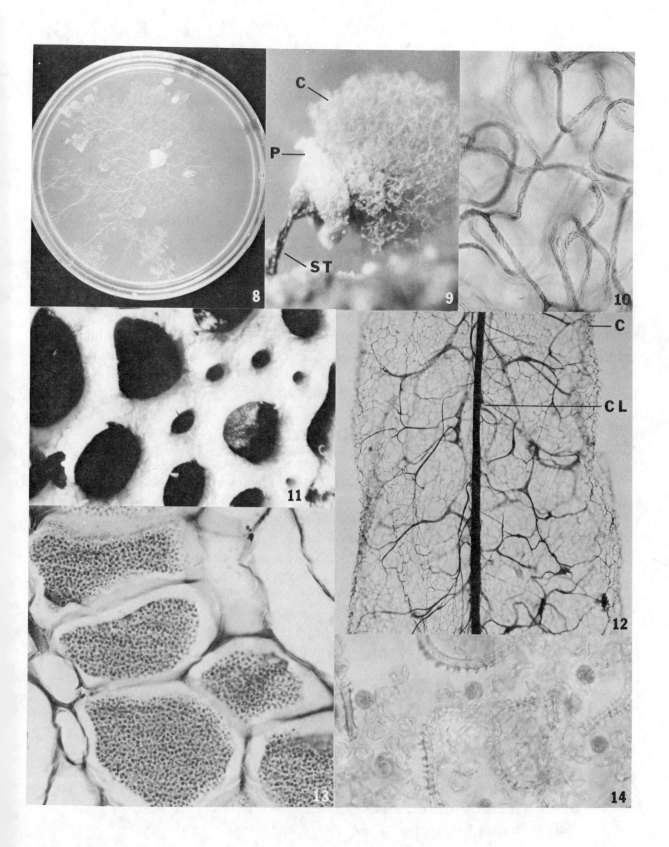

Eumycota

In the Eumycota the assimilative or vegetative thallus has a definite wall and no plasmodia or pseudoplasmodia are formed. Typically the thallus is filamentous, but in some species it is unicellular. The thallus may be septate or aseptate (coenocytic); in septate species the cytoplasm is connected through the septal pores. Thallus cells may be uni- or multi-nucleate. In the majority of true fungi the vegetative thallus is haploid, but occasionally it is diploid. Both sexual and asexual reproduction occur; the subdivisions of Eumycota are separated primarily on the basis of the method of sexual reproduction. Members of the Eumycota occur everywhere, in soils, water, air, and in the tissues of plants and animals, where some cause serious diseases.

Mastigomycotina

The thallus of zoosporic fungi varies from unicellular to mycelial, with the hyphae basically being coenocytic. Septa often form in older hyphae and they always delimit reproductive structures. The hyphae of certain groups are characteristically septate. The mode of sexual reproduction is variable, but results in a zygote which becomes a resting spore and which germinates to form a sporangium. Asexual reproduction is by means of zoospores or non-motile sporangiospores or sporangiola. They occur in nearly every type of habitat.

Chytridiomycetes

The class Chytridiomycetes contains those fungi in which the zoospores have a single, posterior, whiplash flagellum. The thallus varies from unicellular to filamentous and is coenocytic. It may be holocarpic or eucarpic, monocentric or polycentric. The cell wall is composed of chitin and beta-glucan. Most species are minute and occur as parasites on freshwater plants and animals, terrestrial vascular plants, and as saprobes on plant and animal debris. In modern classification the class is separated into orders on the basis of the internal structure of the zoospore.

Chytridiales

In the zoospore of the Chytridiales the ribosomes are enclosed in a membrane-bound cluster, there is a single lateral lipid globule, a rumposome is present, the microtubules lie parallel to each other and extend from the kinetosome to the rumposome, the mitochondria are associated with the lipid globule, the non-functional centriole lies parallel to the kinetosome, and the nucleus is not associated with the kinetosome. They occur primarily in aquatic or semi-aquatic habitats.

Synchytriaceae — Thallus holocarpic and formed exogenous to the zoospore cyst.

Synchytrium endobioticum — This is a holocarpic, endobiotic parasite of potato tubers. Examine prepared slides of this species and study as many stages as possible. You may not find all stages on a single slide. Young thalli, sori, and resting sporangia are most common. The TRIARCH slides are best for the early stages (Fig. 17-19).

Chytridiaceae — Thallus eucarpic, monocentric, with sporangium developed endogenous to zoospore cyst. Sporangia may be operculate or inoperculate.

Cylindrochytridium johnstonii — A monocentric, eucarpic chytrid. Examine the demonstration slide of a sporangium and note the operculum (Fig. 25).

Allochytridium — Make a mount of material in culture and look for sporangia with opercula.

Rhizophydium — This is a monocentric, eucarpic fungus with rhizoids. Mount living material of *Rhizophydium pollinis-pini* in water and observe sporangia and rhizoids. If mature sporangia are available, observe zoospore discharge (Fig. 20-22).

Spizellomycetales

In the zoospore of the Spizellomycetales the ribosomes are dispersed in the cytoplasm, there is a variable number of lipid globules in the anterior, there is no rumposome, the microtubules radiate into the body of the zoospore from the kinetosome, the mitochondria are not associated with the lipid body complex, the non-functional centriole is situated at an angle to the kinetosome, and the nucleus is associated with the kinetosome. These are mainly soil-inhabiting fungi.

Spizellomycetaceae — Thallus eucarpic, monocentric, with the sporangia and resting spore developed endogenous with the zoospore cyst.

Rhizophlyctis — Species in this genus are monocentric and eucarpic, but with an endobiotic apophysis. Mount living material in water and observe sporangia (Fig. 23-24). Watch zoospore discharge if mature sporangia are available.

Olpidiaceae — Sporangium and resting spore developed exogenous to the zoospore cyst. Usually endobiotic.

Rozella — This is an endobiotic, holocarpic parasite of various aquatic fungi. *Rozella allomycis* parasitizes the hyphae and reproductive structures of *Allomyces*. Mount infected *Allomyces* thalli in water and look for young thalli and resting spores of the parasite (Fig. 15-16).

Blastocladiales

The Blastocladiales are characterized by having a well-developed thallus, often hypha-like, on which are borne the reproductive structures. A thick-walled resting sporangium is formed. Both gametes are motile. In the zoospore the ribosomes are membrane-bound and arranged in a nuclear cap and the microtubules radiate into the cytoplasm from the kinetosome.

Blastocladiaceae — Thallus walled, bearing rhizoids, typically with a prominent basal part on which are the reproductive structures.

Allomyces javanicus — This fungus is interesting in that it has an alternation of generations with distinct gametophytic and sporophytic thalli. Examine living material of both of these. Observe the gametangia (male terminal) and watch for escaping gametes. Gamete release is usually preceded by rapid movement of gametes inside the gametangia. Also look for resting sporangia (meiosporangia) and zoosporangia (mitosporangia) (Fig. 26-32).

Monoblepharidales

Members of this order have a delicate, much-branched, hyphal thallus without a well-defined basal cell. Sexual reproduction is oogamous; only the male gamete is motile. In the zoospore the microtubules radiate into the zoospore body from the kinetosome.

Monoblepharidaceae — Zygote not motile, remaining in oogonium or oozing to surface of oogonium, where it encysts.

Monobleparis sp. — Examine the demonstration slide of a thallus and note the antheridium, oogonium and oospore (Fig. 33).

REFERENCES

Barr, D.J.S. 1980. An outline for the reclassification of the Chytridiales, and for a new order, the Spizellomycetales. *Canad. J. Bot.* 58: 2380-2394.

Barr, D.J.S., and V.E. Hadland-Hartmann. 1978. Zoospore ultrastructure in the genus *Rhizophydium* (Chytridiales). *Canad. J. Bot.* 56: 2380-2404.

Emerson, R. 1941. An experimental study of the life cycles and taxonomy of *Allomyces*. *Lloydia* 4:77-144.

Emerson, R., and C.M. Wilson. 1954. Interspecific hybrids and the cytogenetics and cytotaxonomy of *Euallomyces*. *Mycologia* 46: 393-434.

Fuller, M.S. (ed.). 1978. *Lower Fungi in the Laboratory.* Dept. Botany, Univ. Georgia,

Fuller, M.S., and A. Jaworski. (eds.). 1986. *Zoosporic Fungi in Teaching and Research.* Palfrey Contrib. in Botany, No. 3, Dept. of Botany, Univ. Georgia, Athens.

Karling, John S. 1964. *Synchytrium.* Academic Press, New York.

Karling, J.S. 1977. *Chytridiomycetarum Iconographia.* J. Cramer, Vaduz.

Sparrow, F.K., Jr. 1960. *Aquatic Phycomycetes.* 2nd ed. Univ. Michigan Press, Ann Arbor.

FIG. 15. Young thalli of *Rozella allomycis* in vegetative cell of *Allomyces javanicus*. × 1000. FIG. 16. Resting spores of *Rozella allomycis* in vegetative cell of *Allomyces javanicus*. × 1000. FIG. 17-19. *Synchytrium endobioticum* in potato tissue. FIG. 17. Young thalli. × 400. FIG. 18. Sorus. × 400. FIG. 19. Resting spores. × 400. FIG. 20-21. *Rhizophydium pollinis-pini*. FIG. 20. Young thallus. × 1000. FIG. 21. Older thallus with rhizoids. x 1000. FIG. 22-24. *Rhizophlyctis* sp. FIG. 22. Sporangium with apophysis (arrow) and rhizoids. x 1000. FIG. 23. Young thallus with rhizoids. x 1000. FIG. 24. Mature sporangium with two papillae and zoospores. x 1000. FIG. 25. Sporangium with operculum of *Cylindrochytridium johnstonii*. x 1000.

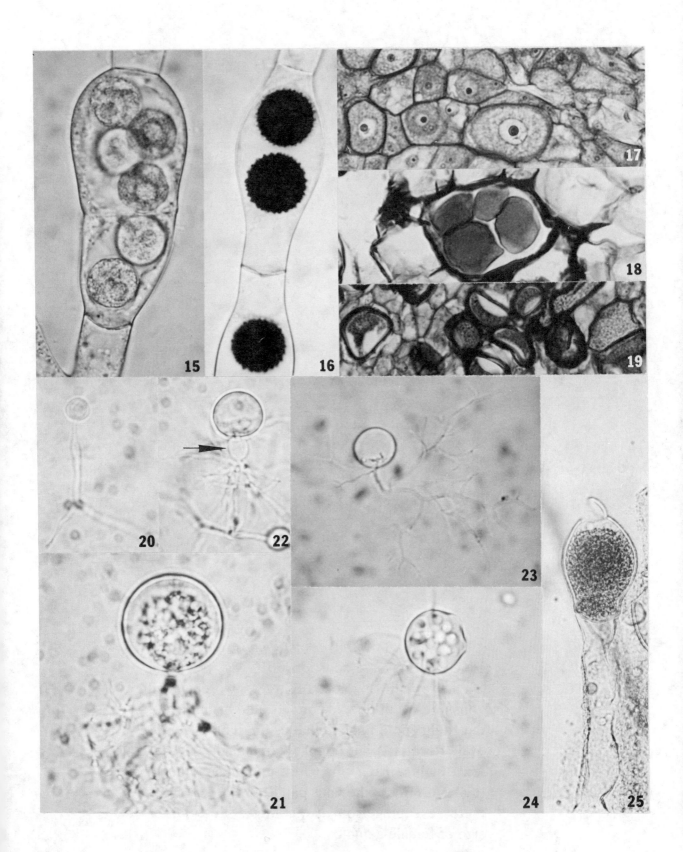

FIG. 26-31. *Allomyces javanicus*. All × 1000. FIG. 26. Encysted zoospore. FIG. 27. Encysted zoospore with rhizoids. FIG. 28. Two encysted zoospores with rhizoids on left, and beginning of development of haploid thallus on right. FIG. 29. Male (terminal) and female gametangia on haploid thallus. FIG. 30. Close-up of gametangia (male terminal) showing papillae. FIG. 31. Gametes escaping from gametangia. Note larger size of female gametes.

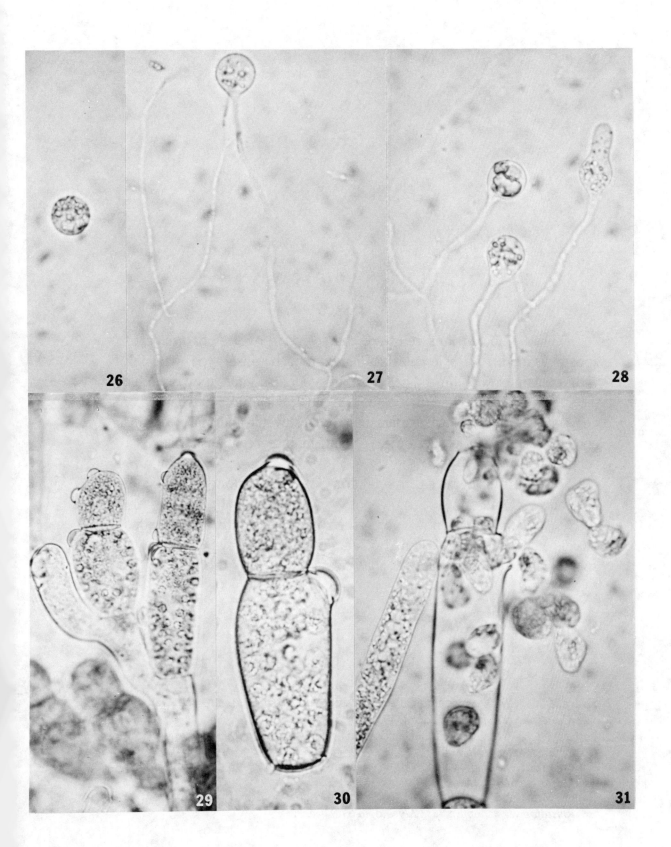

Hyphochytriomycetes

The Hyphochytriomycetes are characterized by the possession of zoospores having a single anterior flagellum of the tinsel type. The cell walls contain both chitin and cellulose. In general structure they resemble the Chytridiomycetes and may be holocarpic or eucarpic, monocentric or polycentric. They occur as parasites on freshwater and marine plants and animals and as saprobes on dead plant and insect remains. There is a single order, the **Hyphochytriales**.

Rhizidiomycetaceae — Thallus eucarpic, monocentric, and epibiotic.

Rhizidiomyces — In this genus the zoospores form in a thin-walled vesicle outside the sporangium. Look at the 1962 paper by M.S. Fuller on sporangial development in *Rhizidiomyces*.

REFERENCES

Fuller, M.S. 1960. Biochemical and microchemical study of the cell walls of *Rhizidiomyces sp. Amer. J. Bot.* 47: 838-842.

Fuller, M.S. 1962. Growth and development of the water mold *Rhizidiomyces* in pure culture. *Amer. J. Bot.* 49: 64-71.

Fuller, M.S., and I. Barshad. 1960. Chitin and cellulose in the cell walls of *Rhizidiomyces sp. Amer. J. Bot.* 47: 105-109.

Fuller, M.S., and R. Reichle. 1965. The zoospore and early development of *Rhizidiomyces apophysatus. Mycologia* 57:946-961.

Plasmodiophoromycetes

In the Plasmodiophoromycetes the thallus is a naked, holocarpic, endobiotic, parasitic plasmodium. The sporangiogenous plasmodium forms sporangia in the host cells and cleaves into spores which give rise to zoospores. The zoospores are biflagellate, with one short, smooth flagellum directed anteriorly and a long, whiplash flagellum directed posteriorly. Thick-walled resting sporangia (cysts) are also formed in the host tissue. As obligate parasites these fungi inhabit algae, fungi, and angiosperms, where they usually occur in the roots.

There is a single order, **Plasmodiophorales**, with one family, **Plasmodiophoraceae**.

Plasmodiophora — In this genus resting sporangia (cysts) are present and they are not united into clusters, but lie free in the host cell. Examine prepared slides of *Plasmodiophora brassicae*, cause of the club root of cabbage, and observe plasmodia and resting sporangia (Fig. 13).

REFERENCES

Aist, J.R., and P.H. Williams. 1971. The cytology and kinetics of cabbage root hair penetration by *Plasmodiophora brassicae. Canad. J. Bot.* 49: 2023-2034.

Garber, R.C., and J.R. Aist. 1979. The ultrastructure of mitosis in *Plasmodiophora brassicae* (Plasmodiophorales). *J. Cell Sci.* 40: 89-110.

Ingram, D.S., and I.C. Tommerup. 1972. The life history of *Plasmodiophora brassicae* Woron. *Proc. R. Soc. London, Ser. B.*, 180: 103-112.

Karling, J.S. 1968. *The Plasmodiophorales*. 2nd ed. Hafner Pub. Co., N.Y.

Kole, A.P., and A.J. Gielink. 1963. The significance of the zoosporangial stage in the life cycle of the Plasmodiophorales. *Neth. J. Plant Path.* 69: 258-262.

Macfarlane, I. 1970. Germination of resting spores of *Plasmodiophora brassicae. Trans. Brit. Mycol. Soc.* 55: 97-112.

Tommerup, I.C., and D.S. Ingram. 1971. The life-cycle of *Plasmodiophora brassicae* Woron. in *Brassica* tissue cultures and in intact roots. *New Phytol.* 70: 327-332.

Williams, P.H., S.J. Aist, and J.R. Aist. 1971. Response of cabbage root hairs to infection by *Plasmodiophora brassicae. Canad. J. Bot.* 49: 41-47.

Williams, P.H., and Y.B. Yukawa. 1967. Ultrastructural studies on the host-parasite relations of *Plasmodiophora brassicae. Phytopathology* 57: 682-687.

Oomycetes

The Oomycetes are characterized by having biflagellate zoospores, each with one tinsel flagellum directed forward and one whiplash flagellum directed backward. Sexual reproduction is oogamous. In most species the thallus is a well-developed, coenocytic mycelium. Current evidence indicates that the oomycete vegetative thallus is diploid. The Oomycetes are typically aquatic, but they are also common in soils. Some species have evolved as parasites of terrestrial plants but they remain dependent upon free water for reproduction. The cell wall is composed of cellulose and beta-glucan.

Saprolegniales

This order includes those Oomycetes having the zoospores always formed within the sporangium; the hyphae in eucarpic forms are not constricted.

Saprolegniaceae — Thallus is filamentous, eucarpic, and never septate. Zoosporangia are delimited by a basal septum. Female oogonia and oospores are always distinct. The family contains many common water molds, some of which are economically important.

Achlya — In *Achlya* at least some zoospores encyst within or at the mouth of the zoosporangium upon discharge. Make a mount of living material of *Achlya americana* and examine for oogonia, oospheres (or oospores) (Fig. 34), and antheridia.

Saprolegnia — In this genus the zoospores swim away immediately upon discharge. Mount and examine living material of *Saprolegnia ferax* for zoosporangia (Fig. 35-38).

Peronosporales

In the Peronosporales the biflagellate, reniform zoospores are formed either within the sporangium or in an evanescent vesicle that arises from the sporangium. In some species the sporangium may germinate by means of a germ tube. The thallus is eucarpic.

Pythiaceae — Saprobic or non-obligate parasites. The sporangiophores are usually undifferentiated from the mycelium, indeterminate.

Pythium — In *Pythium* the zoospores are differentiated in a vesicle that arises from the sporangium. Make a mount from a culture of *Pythium ultimum* and look for sporangia, antheridia, and oogonia with oospheres (Fig. 42). If available, watch zoospore discharge in *Pythium dissotocum* under the microscope (Fig. 39-41).

Pythium aphanidermatum — Examine the demonstration showing the attraction of zoospores of this pathogenic species to freshly cut, living, tomato roots.

Phytophthora — In *Phytophthora* the zoospores form fully within the sporangium before they emerge. Examine prepared slides of *Phytophthora infestans* for sporangiophores and sporangia (Plate III).

Phytophthora palmivora — Mount some deciduous sporangia in a drop of water and cover with a cover glass. Observe zoospore maturation and discharge under the compound microscope.

Peronosporaceae — Obligate parasites with branched sporangiophores of determinate growth, producing sporangia singly.

Plasmopara — In this genus the sporangiophore branches rather randomly and the branches are straight and stiff. The branchlets are short and truncate. Examine prepared slides of *Plasmopara viticola* for sporangiophores and sporangia (Plate III).

Plasmopara geranii — Mount some sporangiophores from dried material and note the shape characteristic for the genus.

Peronospora — The sporangiophores in *Peronospora* are branched, with the branches being curved and tapered (Plate III). Mount some sporangiophores from dried material of *Peronospora manshurica* or other species and note the characteristic shape.

Peronospora effusa, P. geranii — Examine prepared slides of one of these species and look for sporangiophores and sporangia.

Albuginaceae — Obligate parasites with unbranched, clavate sporangiophores, each bearing a chain of deciduous sporangia in subepidermal sori.

Albugo — Species in this genus form chains of sporangia on subepidermal, club-shaped sporangiophores. Examine prepared slides of *Albugo bliti* (Fig. 43) and *Albugo candida* (= *Cystopus candidus*) for oospheres, oospores, and sporangia (Fig. 44-46). Note the difference in ornamentation of the oospores in the two species. Not all stages will be found on one slide. Make a mount of sporangia from dried material of *Albugo* sp. on morning glory and examine under the microscope.

REFERENCES

Bryant, T.R., and K.L. Howard. 1969. Meiosis in the Oomycetes: I. A microspectrophotometric analysis of nuclear deoxyribonucleic acid in *Saprolegnia terrestris*. *Amer. J. Bot.* 56: 1075-1083.

Campbell, W.A., and F.F. Hendrix, Jr. 1967. A new heterothallic *Pythium* from southern United States. *Mycologia* 59: 274-278.

Coker, W.C. 1923. *The Saprolegniaceae, with notes on other water molds*. Univ. N. Carolina Press, Chapel Hill.

Coker, W.C., and V.D. Matthews. 1937. Blastocladiales, Monoblepharidales, Saprolegniales. *N. Amer. Flora* 2(1): 1-76.

Erwin, D.C., S. Bartnicki-Garcia, and P.H. Tsao. (eds.). 1983. *Phytophthora, its Biology, Taxonomy, Ecology, and Pathology*. Amer. Phytopathol. Soc., St. Paul.

Flanagan, P.W. 1970. Meiosis and mitosis in Saprolegniaceae. *Canad. J. Bot.* 48: 2069-2076.

Galindo, A.J., and M.E. Gallegly. 1960. The nature of sexuality in *Phytophthora infestans*. *Phytopathology* 50: 123-128.

Hepting, G.H., T.S. Buchanan, and L.W.R. Jackson. 1945. Little leaf disease of pine. *U.S.D.A. Circ.* 716: 1-15.

Hoch, H.C., and J.E. Mitchell. 1972. The ultrastructure of *Aphanomyces euteiches* during asexual spore formation. *Phytopathology* 62: 149-160.

Howard, K.L., and T.R. Bryant. 1971. Meiosis in the Oomycetes: II. A microspectrophotometric analysis of DNA in *Apodachlya brachynema*. *Mycologia* 68: 58-68.

Johnson, T.W., Jr. 1956. *The genus Achlya: Morphology and taxonomy.* Univ. Michigan Press, Ann Arbor.

Marx, D.H. and W.C. Bryan. 1969. Effect of soil bacteria on the mode of infection of pine roots by *Phytophthora cinnamomi*. *Phytopathology* 59: 614-619.

Middleton, J.T. 1943. The taxonomy, host range and geographic distribution of the genus *Pythium*. *Mem. Torrey Bot. Club* 20: (1): 1-171.

Newhook, F.J., G.M. Waterhouse, and D.J. Stamps. 1978. Tabular key to the species of *Phytophthora* de Bary. *CMI Mycol. Pap.* 143: 1-20.

Papa, K.E., W.A. Campbell, and F.F. Hendrix, Jr. 1967. Sexuality in *Pythium sylvaticum*: Heterothallism. *Mycologia* 59: 589-595.

Scott, W.W. 1961. A monograph of the genus *Aphanomyces. Va. Agr. Exp. Sta. Tech. Bull.* 151: 1-95.

Vujicic, R. 1971. An ultrastructural study of sexual reproduction in *Phytophthora palmivora. Trans. Brit. Mycol. Soc.* 57: 525-530.

Waterhouse, G.M. 1967. Key to *Pythium* Pringsheim. CMI *Mycol. Papers*, No. 109: 1-15.

Waterhouse, G.M. 1967. The genus *Pythium*. CMI *Mycol. Papers*, No. 110: 1-71.

Waterhouse, G.M. 1970. The genus *Phytophthora* de Bary. CMI *Mycol. Papers*, No. 122: 1-59.

FIG. 32. Resting sporangia of *Allomyces javanicus* on diploid thallus. x 500. FIG. 33. Oogonium of *Monoblepharis* sp. with laterally attached antheridium on right and mature oospore at apex. x 800. FIG. 34. Oogonium of *Achyla americana* with oospores. x 400. FIG. 35-38. *Saprolegnia ferax* (phase contrast). All x 320. FIG. 35. Young zoosporangium. FIG. 36. Mature zoosporangium with zoospores. FIG. 37. Beginning of zoospore discharge. FIG. 38. Nearly empty zoosporangium and zoospores. FIG. 39-41. *Pythium dissotocum* (phase contrast). All x 500. FIG. 39. Young sporangium. FIG. 40. Sporangial vesicle. FIG. 41. Mature zoospores in vesicle. FIG. 42. Oogonia of *Pythium ultimum* with oospores. x 750. FIG. 43. Section through host leaf (HL) with broken epidermis (EP) showing layer of sporangiophores with sporangia, and oospore of *Albugo bliti* (arrow) x 250. FIG. 44-46. *Albugo candida*. FIG. 44. Section through oospore in host tissue. x 760. FIG. 45. Close-up of sporangiophores with chains of sporangia. x 400. FIG. 46. Surface ornamentation of oospore. x 760. FIG. 35-41. courtesy M.S. Fuller.

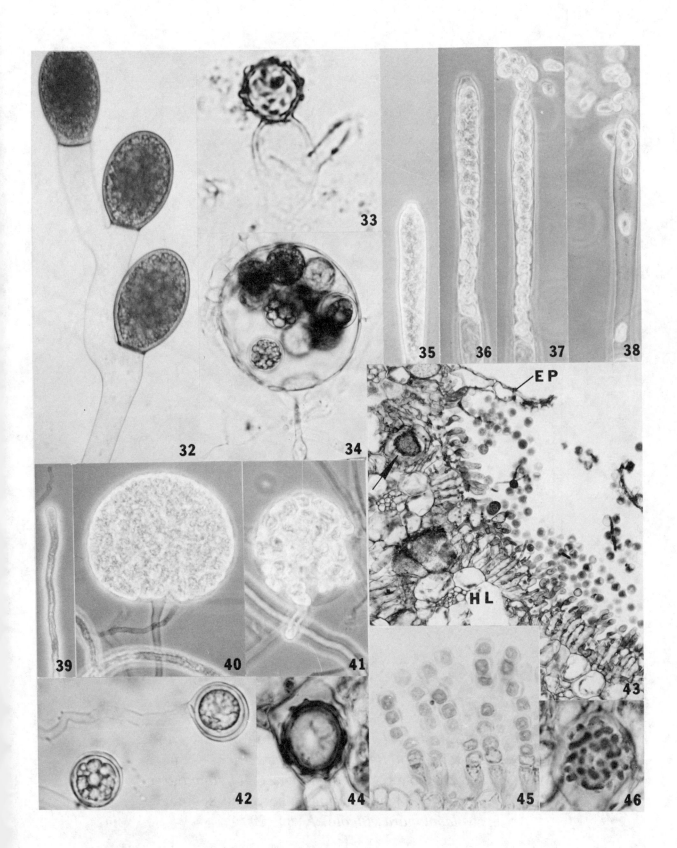

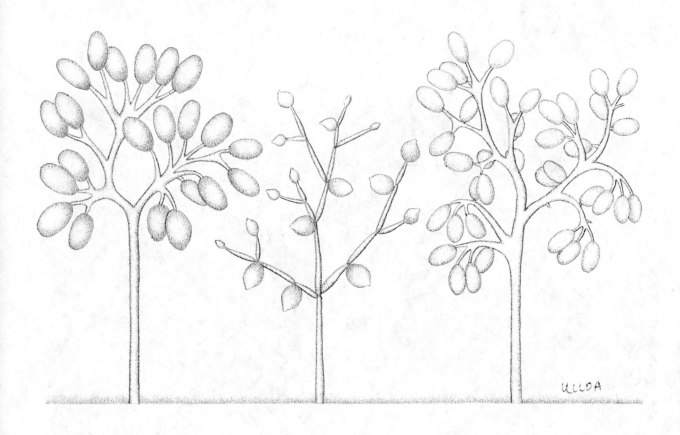

PLATE III. Left, sporangiophore and sporangia of *Peronospora parasitica*. Center, sporangiophore and sporangia of *Phytophthora infestans*. Right, sporangiophore and sporangia of *Plasmopara viticola*. A11 × 200.

Zygomycotina

In the Zygomycotina no motile spores are produced. Sexual reproduction results in the formation of a thick-walled zygospore in a zygosporangium.

Zygomycetes

The Zygomycetes typically reproduce asexually by means of non-motile sporangiospores or sporangiola. Sexual reproduction is by the fusion of gametangia to form a thick-walled zygosporangium. In most species a well developed mycelium is formed, but in parasitic forms the thallus is often much reduced. The cell wall is composed of chitin and chitosan. Members of the Zygomycetes are typically saprobic, but a number of species are parasites of man, other animals, fungi, and plants. They occur in a wide variety of habitats.

Mucorales

The Mucorales characteristically produce a profuse, well developed mycelium that is fast-growing. Most species are saprobic. The Mucorales reproduce sexually by means of zygospores produced in zygosporangia and they may be either homo- or hetero-thallic. Asexual reproduction is by means of sporangiospores, or sporangiola. Members of the Mucorales are among the most common of fungi and are readily isolated from a wide variety of habitats.

Mucoraceae — Sporangia thin-walled and columellate.

Mucor — In *Mucor* there are no rhizoids or stolons, and the sporangium lacks an apophysis (Plate IV). Nearly all species are heterothallic, but if homothallic the suspensors are of approximately equal size. Make a mount in water of sporangia of *Mucor hiemalis* and note the columella and sporangiospores. The sporangium wall is fragile and breaks away easily, leaving the columella exposed. The columella in intact sporangia can be best observed in young material lightly stained with cotton blue. If available, examine mated + and − strains under the dissecting microscope and observe stages in zygosporangium formation. Mount some mature zygosporangia in water and note the surface ornamentation under the microscope (Fig. 47-50).

Phycomyces — In *Phycomyces* the sporangia are borne on very long, unbranched sporangiophores over 8 cm tall, and which have a metallic luster. Observe colonies of *Phycomyces nitens* growing in large test tubes. Examine plates of mated + and − strains under the dissecting microscope and observe the tong-shaped gametangia and stages of zygosporangium formation. The suspensors bear branched, finger-like projections (Plate IV, Fig. 54 and 56).

Rhizopus — Colonies of *Rhizopus* produce rhizoids and stolons and the sporangia have an apophysis. Mount sporangiophores of *Rhizopus nigricans* in water and observe sporangia, spores, columella, stolons, and rhizoids. The columella is best seen in young material. If available, examine mated + and − strains under the dissecting microscope and observe stages in zygosporangium formation (Plate IV, Fig. 51-53).

Zygorhynchus — This genus resembles *Mucor*, but the species are all homothallic, with unequal suspensors. Make a mount of zygosporangia of *Zygorhynchus vuilleminii* in water and observe under the microscope (Fig. 55).

Pilobolaceae — Sporangium columellate, with a thick, cutinized wall; forcibly discharged.

Pilobolus — In *Pilobolus* the sporangium is borne on a swollen subsporangial vesicle. Examine fresh material under the dissecting microscope and observe sporangial discharge. Note the phototropism of the sporangiophores. Mount some sporangiophores in water and observe the sporangia and vesicle (Plate IV).

Thamnidiaceae — Sporangia, when present, columellate; sporangiola borne on same sporangiophore as the sporangium.

Thamnidium — In *Thamnidium* the branches bearing the sporangiola are dichotomously branched. Make a mount of *Thamnidium elegans* in water and observe the sporangiola under the microscope (Plate IV, Fig. 59).

Choanephoraceae — Sporangia collumellate, with persistent walls which break open in halves. Sporangiola usually also present, borne on distinct stalks. The sporangiospores are dark, with tufts of bristles at their ends.

Blakeslea — This genus has sporangia plus sporangiola. Make a mount from a culture of *Blakeslea trispora* and examine under the microscope (Plate V).

Choanephora — This genus differs from *Blakeslea* in having one-celled sporangiola in addition to the sporangia. Make a mount from a culture of *Choanephora cucurbitarum* and examine under the microscope (Plate V).

Cunninghamellaceae — No sporangia formed; only one-celled sporangiola present.

Cunninghammella — In this genus the sporangiophores are branched and indefinite in length, with the branches ending in fertile vesicles. Mount sporangiophores of *Cunninghamella echinulata* in water and observe under the microscope (Plate V).

Syncephalastraceae — Merosporangia formed; saprobic.

Syncephalastrum — The merosporangia in this genus are borne on swollen tips of the sporangiospores. Make a mount of sporangiospores of *Syncephalastrum racemosum* and observe merosporangia under the microscope (Plate V, Fig. 57-58).

Kickxellaceae — Sporocladia formed: vegetative mycelium septate.

Coemansia — In this genus the sporangiola arise from pseudophialides arranged transversely on the lower surface of septate sporocladia. Mount a portion of a colony in water and observe sporocladia, pseudophialides, and sporangiola under the microscope (Plate V).

Entomophthorales

This is a small order of mycelial fungi that parasitize animals such as insects, or they occur saprobically in soil and on dung. Asexual reproduction is by means of spores that are often forcibly discharged. Sexual reproduction is by means of zygospores. There is one family, the **Entomophthoraceae.**

Entomophthora — In *Entomophthora* the spores are smooth, produced outside the host, and are forcibly discharged. Make a mount in water from a culture of this fungus and observe sporogenous cells and spores. Examine prepared slides (as *Empusa muscae*) of sections through a fly infected with this fungus and observe the sporogenous cells and spores and their relation to the host (Fig. 60-61).

REFERENCES

Benjamin, C.R., and C.W. Hesseltine. 1957. The genus *Actinomucor*. *Mycologia* 49: 240-249.

Benjamin, R.K. 1958. Sexuality in the Kickxellaceae. *Aliso* 4: 149-169.

Benjamin, R.K. 1959. The merosporangiferous Mucorales. *Aliso* 4: 321-433.

Benjamin, R.K. 1961. Addenda to "The merosporangiferous Mucorales." *Aliso* 5: 11-19.

Benjamin, R.K. 1963. Addenda to "The merosporangiferous Mucorales." II. *Aliso* 5: 273-288.

Benjamin, R.K. 1966. The merosporangium. *Mycologia* 58: 1-42.

Bracker, C.E. 1966. Ultrastructural aspects of sporangiospore formation in *Gilbertella persicaria*. In M.F. Madelin, ed., *The Fungus Spore*, pp. 39-55. Butterworths, London.

Cook, R.C., and B.E.S. Godfrey. 1964. A key to the nematode-destroying fungi. *Trans. Brit. Mycol. Soc.* 47: 61-74.

Gustafsson, M. 1965. On species of the genus *Entomophthora* Fres. in Sweden. I. Classification and distribution. *Lantbrukshogskolans Ann.* 31: 103-212.

Gustafsson, M. 1965. On species of the genus *Entomophthora* Fres. in Sweden. II. Cultivation and physiology. *Lantbrukshogskolans Ann.* 31: 405-457.

Gustafsson, M. 1969. On species of the genus *Entomophthora* Fres. in Sweden. III. Possibility of usage in biological control. *Lantbrukshogskolans Ann.* 35: 235-274.

Hesseltine, C.W. 1954. The section Genevensis of the genus *Mucor*. *Mycologia* 46: 358-366.

Hesseltine, C.W. 1955. Genera of the Mucorales with notes on their synonymy. *Mycologia* 47: 344-363.

Hesseltine, C.W., and P. Anderson. 1956. The genus *Thamnidium* and a study of the formation of its zygospores. *Amer. J. Bot.* 43: 696-703.

Hesseltine, C.W., and J.J. Ellis. 1961. Notes on Mucorales, especially *Absidia*. *Mycologia* 53: 405-426.

Hesseltine, C.W., and D.I. Fennell. 1955. The genus *Circinella*. *Mycologia* 47: 193- 212.

O'Donnell, K.L. 1979. *Zygomycetes in Culture*. Dept. of Botany, Univ. Georgia, Athens.

Pady, S.M., C.L. Kramer, D.L. Long, and T.D. McBride. 1971. Spore discharge in *Entomophthora grylli*. *Ann. Appl. Biol.* 67: 145-151.

Robinow, C.F. 1957. The structure and behavior of the nuclei in spores and growing hyphae of Mucorales. II. *Phycomyces blakesleeanus*. *Canad. J. Microbiol.* 3: 791-798.

Schipper, M.A.A. 1978. On certain species of *Mucor* with a key to all accepted species. *Stud. Mycol.* 17: 1-52.

Schipper, M.A.A. 1984. A revision of the genus *Rhizopus*. I. The *Rhizopus stolonifer*-group and *Rhizopus oryzae*. *Stud. Mycol.* 25: 1-19.

Zycha, H., and R. Siepmann. 1969. *Mucorales*. J. Cramer, Lehre.

Trichomycetes

The Trichomycetes are obligate symbionts which live in the gut of arthropods or are attached externally to the exoskeleton. Their thalli are unbranched and coenocytic or branched and septate. Asexual reproduction is by means of trichospores, sporangiospores, arthrospores, or amoeboid cells.

Harpellales

In the Harpellales the spores (trichospores) are produced exogenously, and usually bear one or more long, fine appendages. The thalli are branched or unbranched.

Genistellaceae— Thallus branched, attached to hindgut lining.

Smittium — In *Smittium* the trichospores are straight, ellipsoidal, with a collar, and they bear more than one appendage. Make a mount in water of thalli and examine the thallus and trichospores under the microscope (Fig. 62-63).

REFERENCES

Lichtwardt, R.W. 1973. The Trichomycetes: what are their relationships. *Mycologia* 65:1-20.

Manier, J.F., and R.W. Lichtwardt. 1968. Revision de la systematique des Trichomycetes. *Ann. Sci. Nat. Bot.*, Ser. 12, 519-532.

PLATE IV. Far left, Sporangiophores and sporangia of *Phycomyces nitens*, with zygosporangium at base. Sporangia x 10, zygosporangium x 35. Center, Sporangiophore of *Thamnidium elegans* with terminal sporangium and lateral sporangiola. x 100. Left center, Sporangiophores and sporangia of *Rhizopus nigricans*, with zygosporangia at base. Sporangia x 30, zygosporangia x 100. Right center, Sporangiophores and sporangia of *Mucor rouxianus*. x 25. Far right, Sporangiophores, subsporangial vesicle, and sporangia of *Pilobolus kleinii*. x 30.

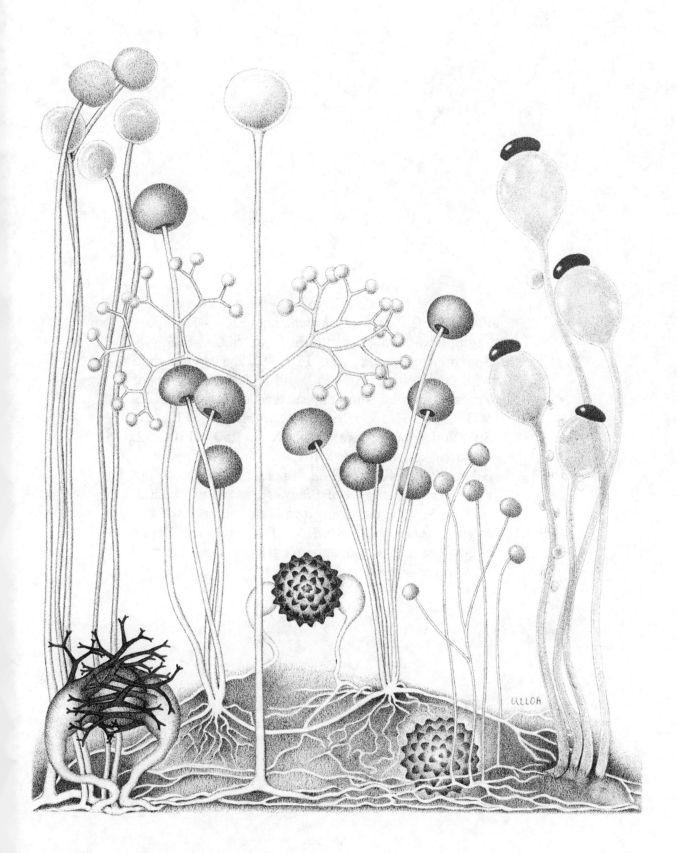

FIG. 47-50. *Mucor hiemalis*. FIG. 47. Sporangiophore and sporangium. x 600. FIG. 48. Sporangiospores. x 400. FIG. 49. Zygosporangium with suspensors. x 250. FIG. 50. Close-up of zygosporangium showing surface ornamentation. x 600. FIG. 51-53. *Rhizopus nigricans*. FIG. 51. Sporangium on sporangiophore. x 250. FIG. 52. Mature sporangium with sporangiospores and columella. x 400. FIG 53. Sporangiospores. x 400. FIG. 54. Gametangia of *Phycomyces nitens*. x 200. FIG. 55. Zygosporangium and suspensor of *Zygorhynchus vuillemini*. x 800. FIG. 56. Young zygosporangium and appendages on suspensors of *Phycomyces nitens*. x 133.

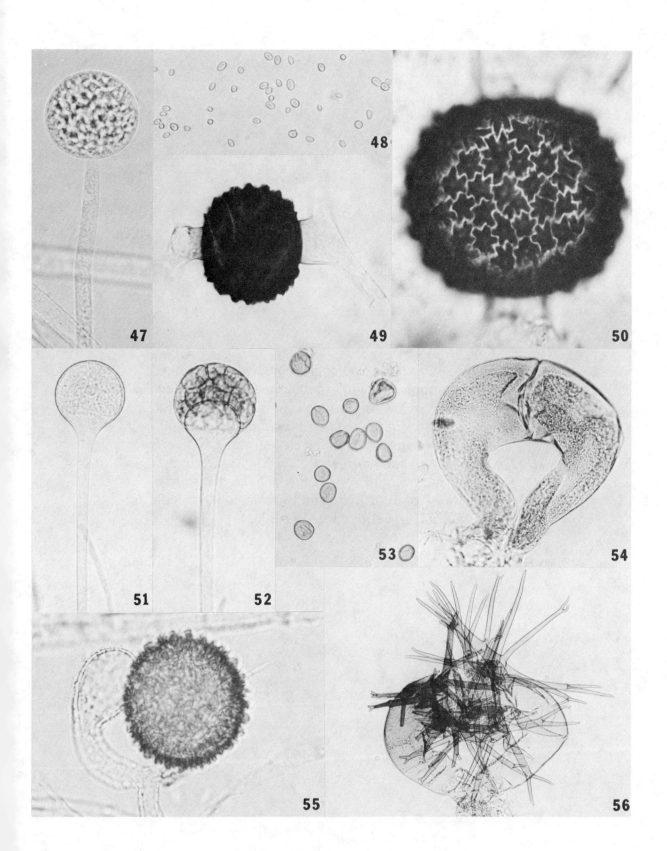

PLATE V. Far left. Sporangiophores with sporangia and sporangiophores with sporangiola of *Blakeslea trispora*. x 300. Left center, Sporangiophore with sporangiola of *Cunninghamella echinulata*. x 200. Center, Sporangiophore with sporangiola of *Choanephora cucurbitarum*. x 100. Right center, Sporangiophore with merosporangia of *Syncephalastrum racemosum*. x 100. Far right, Sporangiophore with sporocladia of *Coemansia reversa*. x 100.

FIG. 57-58. *Syncephalastrum racemosum.* FIG. 57. Sporangiophore with head of merosporangia. x 200. FIG. 58. Merosporangium containing spores, and loose spores. x 600. FIG. 59. Sporangiophore branch with sporangiola of *Thamnidium elegans.* x 320. FIG. 60-61. *Entomophthora muscae.* FIG. 60. Sporogenous cell with spore. x 400. FIG. 61. Section through house fly showing layer of sporogenous cells with spores. x 120. FIG. 62-63. *Smittium* sp. FIG. 62. Close-up of two trichospores attached to thallus. x 400. FIG. 63. Thallus with trichospores. x 100.

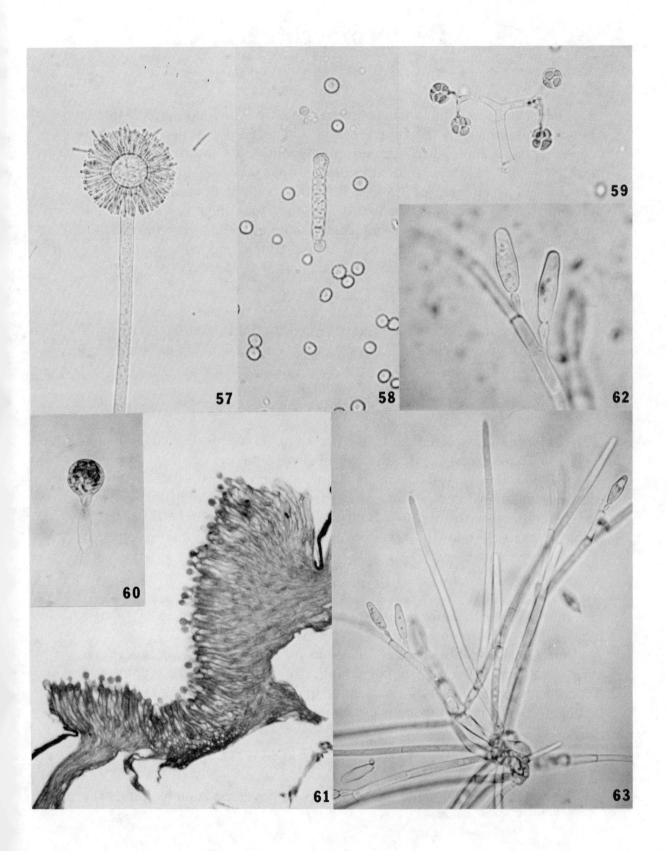

Deuteromycotina

The Deuteromycotina includes those fungi for which no perfect state is known. In practice only the imperfect states of ascomycetes and basidiomycetes are placed here, since the asexual phycomycetes and rusts are recognizable as such and are classified in their proper classes. Most deuteromycetes have a well developed, septate mycelium with distinct conidiophores, but some have a unicellular thallus. A few imperfect fungi lack conidia and form only sclerotia. The deuteromycetes commonly occur as saprobes but some are parasitic on both plants and animals, including man.

Blastomycetes

The Blastomycetes are characterized by budding, yeast-like cells, with or without pseudomycelium. True mycelium is either lacking or is not well developed.

Sporobolomycetales

This order is characterized by the presence of forcibly discharged ballistospores. There is one family, the **Sporobolomycetaceae.**
Sporobolomyces — The thallus is unicellular and forms ballistospores. Mount a small portion of a colony in water and observe the thallus cell and the spore attachment (Fig. 64). Ballistospore discharge can be observed in culture by using the low power objective of the compound microscope.

Cryptococcales

In this order the spores are not forcibly discharged. There is a single family, the **Cryptococcaceae.**
Rhodotorula — Members of this genus form pink colonies of unicellular cells which reproduce by budding. Mount a small portion of a colony of *Rhodotorula rubra* in water and examine under the microscope. You should be able to observe blastogenous conidium formation (budding) (Fig. 65).

Hyphomycetes

In the Hyphomycetes the mycelium is well developed and budding cells are absent. The conidia are borne on conidiophores which are not in a fruiting body. Some species form only sclerotia. They are placed in a single order, the **Moniliales**.

Agonomycetaceae — Sclerotia present, no conidia formed.

Papulaspora — In this genus the sclerotia consist of compact clusters of small cells. Mount a small portion of a colony in water and observe the sclerotia (bulbils) (Fig. 66).

Sclerotium — In *Sclerotium* the sclerotia are usually relatively large and hard, and are composed of very compact hyphae. Observe sclerotia of *Sclerotium rolfsii* under the dissecting microscope. Mount some sclerotia of *Sclerotium bataticola* in water and observe under the microscope (Fig. 67-68).

Moniliaceae — Conidia and hyphae hyaline or light-colored.

Aspergillus — This genus is characterized by a swollen vesicle at the apex of the conidiophore. Conidia are produced by phialides that are either borne directly on the vesicle (uniseriate) or on metulae borne on the vesicle (biseriate). Mount several conidiophores of *Aspergillus clavatus* and observe the uniseriate phialides on the large clavate vesicle. This is an example of phialogenous conidium development (Fig. 71). Mount several conidiophores of a biseriate species, such as *Aspergillus niger* or *A. flavus*, and note the arrangement of the phialides and metulae (Fig. 69-70). The conidial heads in *Aspergillus* are dry and the conidia are amerospores.

Geotrichum — The hyaline mycelium in *Geotrichum* fragments into arthrospores with truncate ends. Mount a small portion of a colony in water and observe fragmenting hyphae and arthrospores (Fig. 72).

Gliocladium — In *Gliocladium* the phialides are arranged in a compact "brush" or penicillus and the conidia are held together in a slime droplet. *Gliocladium roseum* forms pinkish-white colonies. Observe a colony under the dissecting microscope and look for slime droplets. Mount a portion of a colony in water and observe the arrangement of the phialides and metulae. Conidium development is phialogenous and the conidia are amerospores (Fig. 73).

Penicillium — In *Penicillium* the phialides are also arranged in a penicillus, but the spore head is dry. No slime droplet is formed (Fig. 74). In *Penicillium claviforme* the conidiophores are united into a loose, erect structure, the coremium (synnema). Observe a culture under the dissecting microscope and note the coremia. Mount a coremium in water and examine under the microscope. The conidia are amerospores and are formed phialogenously.

Dematiaceae — Conidia and/or hyphae brown or black.

Alternaria — Both the conidia and hyphae are dark brown in *Alternaria*. The conidia are dictyospores and are formed porogenously. Mount young conidiophores of *Alternaria alternata* in water and observe the method of conidium formation. Also mount mature conidia and look at the pattern of septation (Fig. 75-77).

Bipolaris — The conidia and hyphae in *Bipolaris* are dark brown; the conidia are large phragmospores and are formed porogenously on the apex of erect conidiophores. After formation of the conidium the conidiophore elongates sympodially and produces another conidium at the apex. The conidia germinate by one germ tube at each end. Examine cultures of *Bipolaris maydis* under the dissecting microscope, then mount some material in water and examine under the microscope (Fig. 81). This species was formerly placed in the genus *Helminthosporium*.

Curvularia — The conidia in *Curvularia* are dark brown phragmospores. The central cells are swollen and the conidia are often curved; the end cells are usually lighter in color. Mount some young conidiophores in water and observe the porogenous mode of conidium development under the microscope. Mount and observe mature conidia (Fig. 78-79).

Thielaviopsis — This genus produces both dark-brown aleuriospores and hyaline endoconidia with truncate ends. Mount a portion of a colony of *Thielaviopsis basicola* in water and look for brown phialides with endoconidia and phragmosporous aleuriospores (Fig. 82-84). Another name for this fungus is *Chalara elegans*.

Helicosporium — The conidia in this genus are hyaline, helically coiled, and septate. Make a mount from a culture or examine conidia on a demonstration slide under the microscope (Fig. 85).

Orbimyces — This genus produces staurospores; the conidium has an inflated central cell with four radiating arms. Examine conidia under the microscope, either from culture or demonstration slide (Fig. 86).

Stilbellaceae — Conidiophores united into synnemata.

Dendrostilbella — Green amerospores are formed in a slime droplet at the apex of a synnema in *Dendrostilbella*. Observe synnemata in culture under the dissecting microscope, then mount some synnemata in water and observe under the microscope (Fig. 87).

Tuberculariaceae — Conidiophores borne on a sporodochium.

Epicoccum — The conidia in *Epicoccum* are dark-brown dictyospores that are borne in clusters. Mount young mycelium of *Epicoccum nigrum* in water and observe murogenous conidium development under the microscope. Mount and examine mature conidia (Fig. 80).

Myrothecium — In *Myrothecium verrucaria* the conidia form dark-green, slimy masses on the pure white mycelium of the sporodochium. Mount a portion of a sporulating area of a sporodochium in water and look for conidiophores and conidia. The conidia are one-celled and have tapered ends, with a pair of hairlike appendages at one end (Fig. 89). This species has also been placed in the genus *Starkeyomyces*.

Coelomycetes

In the Coelomycetes the conidiophores are borne in a fruiting body.

Melanconiales

Conidiophores are borne in an acervulus. Since an acervulus forms by eruption of the fungus through the host epidermis, it does not form in culture. Consequently, in culture the acervulate fungi resemble Hyphomycetes. They are placed in a single family, the **Melanconiaceae.**

Colletotrichum — This genus typically forms hyphal clumps from which stiff, tapering, dark-brown setae arise. The conidia are hyaline amerospores, and are either curved or straight, depending upon the species. Mount a portion of a colony of *Colletotrichum circinans* in water and look for setae, conidiophores, and conidia under the microscope (Fig. 88 and 90). If available, examine prepared slides showing sections through an acervulus.

Pestalotia — In *Pestalotia* the conidia are brown phragmospores with hyaline end cells bearing setulae. Mount a portion of a colony in water and observe conidiophores and conidia under the microscope (Fig. 95).

Sphaeropsidales
(Phomales)

Conidia are borne in a pycnidium.

Sphaeropsidaceae — Pycnidia dark-colored.

Phoma — In *Phoma* the small, one-celled, hyaline conidia are borne inside an ostiolate, dark-brown pycnidium. The conidia are often extruded from the ostiole in a cirrhus. Examine a colony under the dissecting microscope and look for cirrhi. Mount several pycnidia in water and examine under the microscope (Fig. 91-92).

Septoria — In *Septoria* the conidia are hyaline scolecospores. Examine a prepared slide showing sections through a pycnidium of *Septoria apii* with conidia. Examine *Vaccinium* leaves bearing pycnidia of *Septoria* sp. under the dissecting microscope. Mount several pycnidia in water, crush gently, and examine the scolecosporous conidia under the microscope (Fig. 93-94).

REFERENCES

Barnett, H.L., and B.B. Hunter. 1972. *Illustrated Genera of the Fungi Imperfecti*. 3rd ed. Burgess Publ. Co., Minneapolis.

Barron, G.L. 1968. *The Genera of Hyphomycetes from Soil*. Williams & Wilkins Co., Baltimore.

Bracker, C.E., Jr., and E.E. Butler. 1963. The ultrastructure and development of septa in hyphae of *Rhizoctonia solani*. *Mycologia* 55: 35-58.

Buckley, P.M., T.D. Wyllie, and J.E. DeVay. 1969. Fine structure of conidia and conidium formation in *Verticillium albo-atrum* and *V. nigrescens*. *Mycologia* 61: 240-250.

Carmichael, J.W., W.B. Kendrick, I.L. Conners, and L. Sigler. 1980. *Genera of Hyphomycetes*. Univ. Alberta Press, Edmonton.

Cole, G.T., and W.B. Kendrick. 1969. Conidium ontogeny in hyphomycetes. The annellophores of *Scopulariopsis brevicaulis*. *Canad. J. Bot.* 47: 925-929.

Cole, G.T. and W.B. Kendrick. 1969. Conidium ontogeny of hyphomycetes. The arthrospores of *Oidiodendron* and *Geotrichum*, and the endoarthrospores of *Sporendonema*. *Canad. J. Bot.* 47: 1773-1780.

Cole, G.T., and W.B. Kendrick. 1969. Conidium ontogeny in hyphomycetes. The phialides of *Phialophora, Penicillium*, and *Ceratocystis*. *Canad. J. Bot.* 47: 779-789.

Cole, G.T., and B. Kendrick. 1981. *Biology of Conidial Fungi*. Vols. 1 & 2. Academic Press, New York.

Ellis, M.B. 1971. *Dematiaceous Hyphomycetes*. Commonwealth Mycol. Inst., Kew, England.

Ellis, M.B. 1976. *More Dematiaceous Hyphomycetes*. Commonwealth Mycol. Inst., Kew, England.

Hammill, T.M. 1971. Fine structure of annellophores. I. *Scopulariopsis brevicaulis* and *S. koningii*. *Amer. J. Bot.* 58: 88-97.

Hughes, S.J. 1953. Conidiophores, conidia and classification. *Canad. J. Bot.* 31: 577-659.

Kendrick, B. (ed.). 1971. *Taxonomy of Fungi Imperfecti*. Univ. Toronto Press, Toronto.

Luttrell, E.S. 1963. Taxonomic criteria in *Helminthosporium*. *Mycologia* 55: 643-674.

Luttrell, E.S. 1964. Systematics of *Helminthosporium* and related genera. *Mycologia* 56: 119-132.

Morgan-Jones, G., and B. Kendrick. 1972. Icones genera coelomycetarum III. *Univ. Waterloo Biol. Ser.* 5: 1-42.

Morgan-Jones, G., T.R. Nag Raj, and B. Kendrick. 1972. Icones genera coelomycetarum I. *Univ. Waterloo Biol. Ser.* 3:1-42.

Morgan-Jones, G., T.R. Nag Raj, and B. Kendrick. 1972. Icones genera coelomycetarum II. *Univ. Waterloo Biol. Ser.* 4: 1-40.

Morgan-Jones, G., T.R. Nag Raj, and B. Kendrick. 1972. Icones genera coelomycetarum IV. *Univ. Waterloo Biol. Ser.* 6: 1-42.

Nag Raj, T.R., and B. Kendrick. 1975. *A Monograph of Chalara and Allied Genera*. Wilfrid Laurier Univ. Press, Waterloo.

Subramanian, C.V. 1962. The classification of the Hyphomycetes. *Bull. Bot. Surv. India* 4: 249-259.

Subramanian, C.V. 1983. *Hyphomycetes. Taxonomy and Biology.* Academic Press, New York.

Sutton, B.C. 1980. *The Coelomycetes.* Commonwealth Mycol. Inst., Kew, England.

Tubaki, Keisuke. 1958. Studies on the Japanese hyphomycetes. V. Leaf and stem group with a discussion of the classification of hyphomycetes and their perfect stages. *J. Hattori Bot. Lab.* 20: 142-244.

Tubaki, K. 1963. Taxonomic study of Hyphomycetes. *Ann. Rep. Inst. Ferment. Osaka,* No. 1 (1961-1962): 25-54.

Zachariah, K., and P.C. Fitz-James. 1967. The structures of phialides in *Penicillium claviforme. Canad. J. Microbiol.* 13: 249-256.

FIG. 64. Thallus cell of *Sporobolomyces* sp. with ballistospore on sterigma. × 2000. FIG. 65. Vegetative cells of *Rhodotorula rubra* with blastospores (buds). × 2500. FIG. 66. Bulbils (sclerotia) of *Papulaspora* sp. × 320. FIG. 67. Sclerotium of *Sclerotium bataticola*. × 200. FIG. 68. Sclerotia of *Sclerotium rolfsii*. × 100. FIG. 69-70. *Aspergillus flavus*. FIG. 69. Conidiophore with vesicle bearing metulae and phialides (biseriate). × 600. FIG. 70. Metulae bearing phialides with conidia. × 1000. FIG. 71. Conidiophore with vesicle of *Aspergillus clavatus* bearing phialides (uniseriate). × 560. FIG. 72. Chain of arthrospores on hypha of *Geotrichum candidum* above, with loose arthrospores below. × 1000. FIG. 73. Branched conidiophore of *Gliocladium roseum* bearing phialides and conidia. × 1000. FIG. 74. Conidiophore of *Penicillium* sp. with metulae (arrow), phialides, and conidia. × 1000. FIG. 75-77. *Alternaria alternata*. FIG. 75. Young conidium on conidiophore. × 1200. FIG. 76. Mature conidium. × 1000. FIG. 77. Germinating conidium. × 1000. FIG. 78-79. *Curvularia geniculata*. FIG. 78. Conidium in side view. × 1000. FIG. 79. Conidium in face view. × 1000. FIG. 80. Conidia of *Epicoccum nigrum*. × 1000.

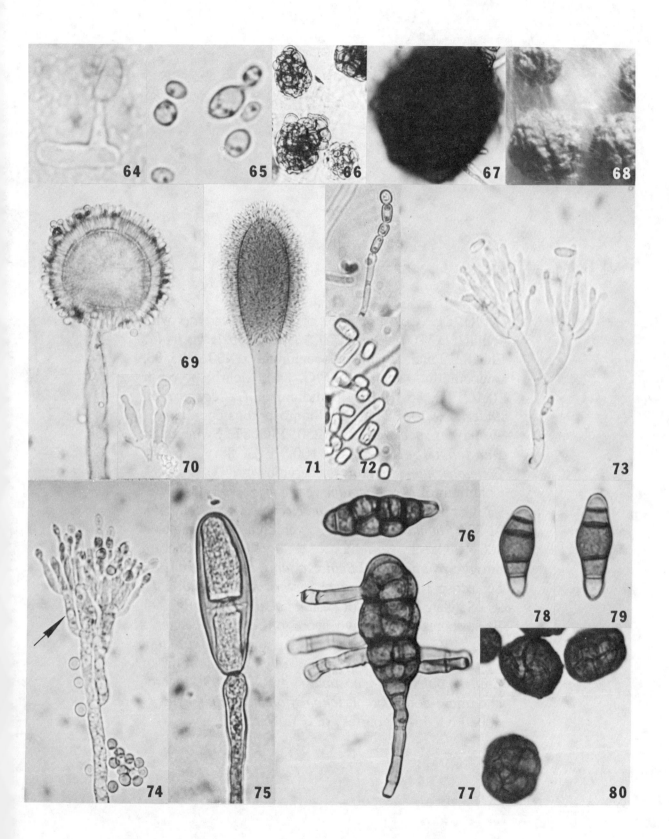

FIG. 81. Conidiophores of *Bipolaris maydis* with conidia. x 760. FIG. 82-84. *Thielaviopsis basicola*. FIG. 82. Phialide with endoconidium. x 400. FIG. 83. Endoconidia. x 1000. FIG. 84. Aleuriospore. x 1000. FIG. 85. Helical conidium of *Helicosporium linderi*. x 700. FIG. 86. Staurosporous conidium of *Orbimyces spectabilis*. x 1250. FIG. 87. Synnema of *Dendrostilbella* sp. x 100. FIG. 88. Setae of *Colletotrichum circinans* in acervulus. x 500. FIG. 89. Conidium of *Myrothecium verrucaria*. x 1000. FIG. 90. Conidia of *Colletotrichum circinans*. x 1760. FIG. 91-92. *Phoma* sp. FIG. 91. Pycnidium with ostiole. x 400. FIG. 92. Conidia. x 1000. FIG. 93. Section through pycnidium of *Septoria apii*. x 400. FIG. 94. Scolecosporous conidium of *Septoria* sp. x 1000. FIG. 95. Setulate conidium of *Pestalotia* sp. x 1000. FIG. 96. Blastospores and chlamydospores of *Candida albicans*. x 400. FIG. 97. Macroconidium of *Microsporum gypseum*. x 400. FIG. 98. Macroconidia of *Epidermophyton floccosum*. x 400. Fig. 99. Macro- and micro-conidia of *Trichophyton mentagrophytes*. x 400. FIG. 96-99 courtesy of G.E. Michaels.

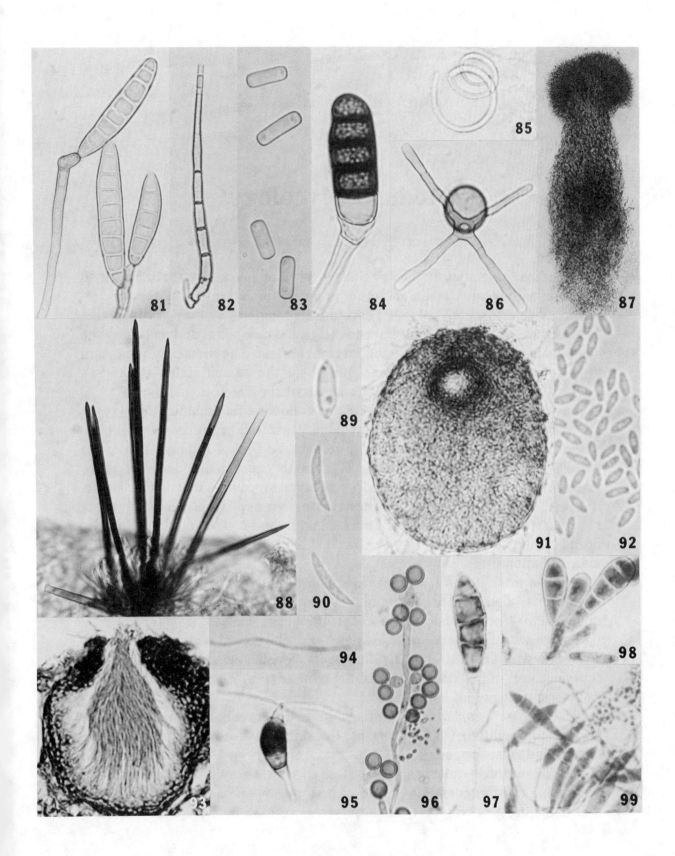

Medical Mycology

Medical mycology is the specialized area of mycology that deals with the fungi pathogenic to man and other animals; the latter is sometimes referred to as veterinary mycology. There are two general classes of fungal diseases, those which are superficial and attack the hair, skin, and nails, and the deep-seated mycoses which attack internal organs. The medically important fungi are mostly deuteromycetes, and where the sexual states are known they usually belong to the Gymnoascaceae of the Plectomycetidae.

Examine demonstration slides of the following pathogenic fungi.

Candida albicans – This is a mycelial yeast that produces hyaline blastospores and clusters of chlamydospores at the tips of hyphae. On agar a white, yeast-like colony is formed. The fungus causes candidiasis in man.

Microsporum gypseum produces hyaline mycelium that forms single, hyaline, spindle-shaped, roughened phragmosporous conidia (macroconidia) on upright conidiophores. Small, one-celled microconidia are borne on the sides of the hyphae (Fig. 97). This fungus causes ringworm in man and animals.

Epidermophyton floccosum has small, broadly clavate, hyaline, phragmosporous, thick-walled macroconidia with smooth walls (Fig. 98). This is the cause of tinea pedis (athlete's foot) in man.

Trichophyton mentagrophytes has large, clavate, phragmosporous macroconidia with smooth, thin, hyaline walls. Small, one-celled, hyaline microconidia are formed on the sides of the hyphae (Fig. 99). This fungus causes ringworm of man and animals.

Histoplasma capsulatum forms white to tan mycelial colonies in culture at room temperature. At 37°C a yeast-like colony is formed. Small, hyaline, microconidia are formed on short, lateral conidiophores. Most characteristic are the large, thick-walled, spherical to pyriform, spiny or tuberculate macroconidia. This fungus is the cause of histoplasmosis, a pulmonary disease that is the most prevalent systemic mycosis of man in the United States.

REFERENCES

Ajello, L., L.K. Georg, W. Kaplan, and L. Kaufman. 1963. *Laboratory Manual for Medical Mycology*. PHSP No. 994. Superintendent of Documents, U.S. Government Printing Office. Washington 25, D.C.

Baker, R.D. 1971. *Human Infection with Fungi. Actinomycetes and Algae*. Springer-Verlag.

Beneke, E.S. 1976. *Scope Monograph on Human Mycoses*. 6th ed. The Upjohn Co., Kalamazoo.

Bulmer, G.S. 1978. *Medical Mycology*. The Upjohn Co., Kalamazoo.

Conant, N.F., D.T. Smith, R.D. Baker, J.L. Callaway, and D.S. Martin. 1971. *Manual of Clinical Mycology*. 31st Ed. W.B. Saunders Co., Philadelphia, Pa.

Emmons, C.W., C.H. Binford, and J.P. Utz. 1970. *Medical Mycology*. 2nd Ed. Lea and Febiger. Philadelphia, Pa.

Hazen, E.S., M.A. Gordon, and F.C. Reed. 1970. *Laboratory Identification of Pathogenic Fungi Simplified*, 3rd Ed. Charles C. Thomas. Springfield, Ill.

Rebell, G., and D. Taplin. 1971. *Dermatophytes. Their Recognition and Identification*. Dermatology Foundation of Miami, 1020 N.W. 16th St., Miami, Fla.

Rippon, J.W. 1982. *Medical Mycology. The Pathogenic Fungi and The Pathogenic Actinomycetes*. 2nd ed. W.B. Saunders Co., Philadelphia.

Ascomycotina

In the Ascomycotina the perfect state spores (ascospores) are formed inside a specialized cell, the ascus. In most ascomycetes there are eight ascospores per ascus, but the number may vary from one to over 1,000, depending upon the species. In most ascomycetes the asci are borne in an ascocarp and the ascospores are forcibly discharged. The majority of ascomycetes produce a well developed, septate mycelium, although a few have a unicellular thallus. In septate species the septum is simple in structure. The ascomycetous cell wall is composed primarily of chitin. Many ascomycetes have abundant conidial or imperfect states in addition to the ascospore. The Ascomycotina constitute the largest group of fungi. They are widespread in distribution and occur in nearly every type of habitat. Many are of considerable economic importance, as food, in industrial processes, and as causes of disease in plants and animals.

Hemiascomycetes

The Hemiascomycetes are ascomycetes in which the asci are borne free (or naked) on the surface of the host or substrate, i.e., they are not enclosed in any kind of fruiting body. No ascogenous hyphae are formed. The thallus is simple in structure; when a mycelium is formed it is often scanty. Although the Hemiascomycetes are widely distributed as a group, they are often associated with ripening fruits and plant exudates.

Endomycetales

The Endomycetales includes those Hemiascomycetes in which two cells fuse to form a zygote, which is then transformed into an ascus.

Dipodascaceae — Mycelial forms with multispored asci that do not proliferate.

Dipodascopsis — The mycelium consists of irregular cells and bears long slender asci which taper toward the apex. The ascospores are small, one-celled, and hyaline. Mount a small amount of material from a colony of *Dipodascopsis uninucleatus* in a drop of water and look for asci of different ages. Note the large number of ascospores in each ascus (Fig. 100). This species was originally included in *Dipodascus*.

Cephalcascaceae — Mycelial fungi in which the asci are borne on an erect ascophore.

Cephaloascus — Mount ascophores of *Cephaloascus albidus* from culture and examine the chains of asci produced at the apex. The ascospores are hat-shaped.

Saccharomycetaceae — Mycelium scanty or lacking, asci 1-8 spored.

Saccharomyces — In *Saccharomyces* the vegetative cells reproduce by budding and fermentation of sugars is vigorous. The ascospores are round. Make a mount from a culture of *Saccharomyces cerevisiae,* the beer and wine yeast, and observe under the microscope. The vegetative cells are oval and vary considerably in size. Observe stages in budding. Also look for asci with ascospores; these are most common in old cultures (Fig. 103).

Saccharomyces kluyveri — Some heterothallic yeasts, such as this species, exhibit the phenomenon of "sexual agglutination," in which the cells precipitate out of suspension when the two mating strains are mixed together. Observe a demonstration of this phenomenon.

Schizosaccharomyces — In this genus vegetative reproduction is solely by fission. Mount a small portion of a colony of *Schizosaccharomyces octosporus* and observe under the microscope. The vegetative cells are hyaline and tend to be elongate. Observe dividing cells. Also look for various stages in cell conjugation and ascus formation. Mature asci are elongate and usually contain eight one-celled ascospores (Fig. 101-102).

Spermophthoraceae — Ascospores spindle or needle-shaped.

Nematospora – This is a mycelial yeast which produces needle-shaped ascospores. Make a mount from a culture of *Nematospora gossypii* and look for asci and ascospores (Fig. 104).

Taphrinales

The Taphrinales includes those Hemiascomycetes in which the ascus arises from binucleate cells that form directly from the mycelium. The asci are erumpent through the host epidermis and are borne free on the host. All are parasitic on higher plants and some cause economically important plant diseases. Usually eight one-celled ascospores are formed, but the number varies from four to many in different species. In other species the ascospores bud within the ascus to form blastospores. In culture the Taphrinales form yeast-like colonies of unicellular budding cells. There is a single family, the **Taphrinaceae,** and one genus.

Taphrina — Examine preserved specimens showing the effects of infection by *Taphrina deformans* on peach (peach leaf curl) and *T. communis* on plum (plum pockets). Examine prepared slides of sections through peach leaves infected with *Taphrina deformans* and look for asci and ascospores (Fig. 105). Mount a small portion of a culture in water and examine under the microscope.

FIG. 100. Ascus and ascospores of *Dipodascopsis uninucleatus* × 800. FIG. 101-102. *Schizosaccharomyces octosporus*. FIG. 101. Vegetative cell dividing by fission. × 1000. FIG. 102. Ascus with eight ascospores. × 2000. FIG. 103. Ascus (arrow) of *Saccharomyces cerevisiae* with four ascospores. Vegetative cells are also present. × 1000. FIG. 104. Ascus and ascospores of *Nematospora gossypii*. × 1000. FIG. 105. Asci with ascospores of *Taphrina deformans* on peach leaf. × 625. FIG. 106-107. *Talaromyces flavus*. FIG. 106. Cleistothecia. × 82. FIG. 107. Ascospores. × 1000. FIG. 108-111. *Emericella rugulosa*. FIG. 108. Cleistothecium. × 230. FIG. 109. Hulle cells. × 400. FIG. 110. Ascus with ascospores. × 1000. FIG. 111. Mature ascospores. × 1000. FIG. 112. Ascus and ascospores of *Gymnoascus reessii*. × 1000. FIG. 113. Section through ascocarp of *Eurotium* sp. showing wall (peridium) and asci with ascospores. × 625. FIG. 114-118. *Ceratocystis fimbriata*. FIG. 114. Mature ascocarp. × 100. FIG. 115. Ascospores clustered at tip of ostiolar neck. × 400. FIG. 116. Chain of cylindrical endoconidia. × 400. FIG. 117. Chain of barrel-shaped endoconidia. × 400. FIG. 118. Hat-shaped ascospores. × 1000. FIG. 119. Synnema of *Pesotum ulmi,* conidial state of *Ceratocystis ulmi.* x 100. FIG. 104 courtesy of L.R. Batra.

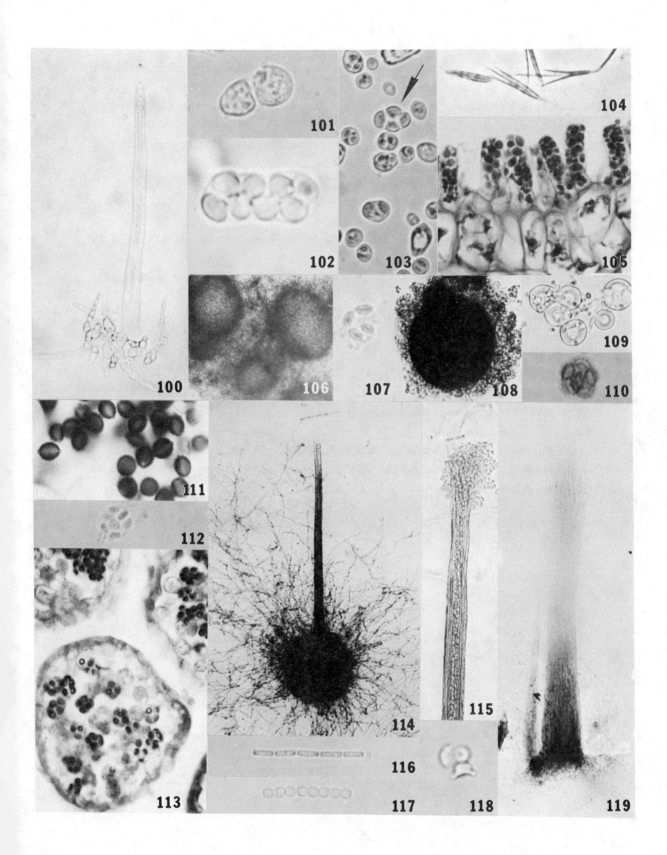

REFERENCES

Batra, L.R. 1959. A comparative morphological and physiological study of the species of *Dipodascus*. *Mycologia* 51: 329-355.

Batra, L.R. 1973. *Nematosporaceae (Hemiascomycetidae): Taxonomy, Pathogenecity, Distribution, and Vector Relations*. U.S. Dept. Agr. Tech. Bull. 1469: 1-71.

Batra, L.R. 1978. Taxonomy and systematics of the Hemiascomycetes (Hemiascomycetidae). In: C.V. Subramanian, *Taxonomy of Fungi*. Univ. Madras, Madras.

Cook, A.H. (ed.) 1958. *The chemistry and biology of yeasts*. Academic Press, New York.

Kramer, C.L. 1960. Morphological development and nuclear behavior in the genus *Taphrina*. *Mycologia* 52: 295-320.

Kurtzman, C.P. 1977. *Cephaloascus albidus*, a new heterothallic yeastlike fungus. Mycologia 69: 547-555.

Lodder, J. (ed.) 1970. *The Yeasts: A Taxonomic Study*. 2nd ed. North Holland Publishing Co., Amsterdam.

Mix, A.J. 1949. A monograph of the genus *Taphrina*. *Univ. Kansas Sci. Bull.* 33: 1-167.

Phaff, H.J., A.W. Miller, and E.M. Mrak. 1966. *The Life of Yeasts*. Harvard University Press, Cambridge.

Rose, A.H., and J.S. Harrison (eds.) The Yeasts. Academic Press, New York.

Vol. 1. 1969. *Biology of Yeasts*.

Vol. 2. 1971. *Physiological and Biochemistry of Yeasts*.

Vol. 3. 1970. *Yeast Technology*.

Taylor, W.S., and R.P. Vine. 1968. *Home Winemaker's Handbook*. Award Books, New York.

Tritton, S.M. 1965. *Tritton's Guide to Better Wine and Beer Making for Beginners*. Farber & Farber, London. In U.S.: Dover Publications, Inc., New York.

Euascomycetes

The asci in the Euascomycetes are unitunicate and they are produced in some kind of fruiting body. In most species the asci are persistent, with a distinct apical apparatus, but some Euascomycetes have evanescent asci. The thallus of the Euascomycetes is a well developed, simple-septate mycelium. The pattern of ascocarp development is ascohymeniaceous, i.e., the ascocarp begins as a more or less coiled hypha which will give rise to the ascogenous system and centrum tissues, and which becomes enveloped by other hyphae to form the ascocarp wall. Most Euascomycete ascocarps are ostiolate. These fungi are widespread in occurrence, but are often associated with plant tissues, either as saprobes or parasites. The class is subdivided on the type of ascocarp that is produced.

Plectomycetidae

In members of this subclass the asci are small, evanescent, and are produced at different levels within the ascocarp. The ascocarp may vary from a loose weft of hyphae bearing asci to a well organized structure with a definite wall. In many plectomycetes the ascocarp is completely enclosed (cleistothecium), but in some it is ostiolate. In general, these fungi are small when compared to other Euascomycetes. Many have profuse conidial states. They are of widespread occurrence, but are often associated with seeds, soils, and they occur as animal parasites.

Eurotiales

The Eurotiales includes those plectomycetes which form their asci in completely enclosed ascocarps (cleistothecia). The ascocarps are small, spherical, and are formed on a well developed mycelium. The asci are globose to sub-globose, usually eight spored, and are evanescent, freeing the ascospores inside the ascocarp. The ascospores are unicellular, frequently ornamented, without germ pores, and are often bivalve in structure. They may be hyaline or colored. The anamorph is usually phialidic.

Eurotiaceae — Cleistothecial wall (peridium) consisting of closely interwoven hyphae.

Emericella — In this genus the ascocarp wall is discrete and composed of flattened polygonal cells. The ascocarp is dark-colored and is surrounded by hyaline, thick-walled hulle cells. The ascospores are red to violet. Examine a culture of *Emericella rugulosa* under the dissecting microscope. Mount several ascocarps and crush gently; look for hulle cells, asci, and the purple-red ascospores (Fig. 108-111). Asci will only be found in young material. Also examine conidiophores of the imperfect state, *Aspergillus rugulosus*.

Eurotium — In *Eurotium* the ascocarps are usually yellow and lack the surrounding hulle cells. The ascocarp wall is one cell layer thick. The ascospores are hyaline. The imperfect state is in *Aspergillus*. Examine prepared slides of sections through ascocarps of *Eurotium* sp. and note the ascocarp wall and the scattered arrangement of the asci (Fig. 113).

Talaromyces — In this genus the ascocarp wall is composed of interwoven hyphae and the ascocarps are usually white or yellow. The asci are formed in short chains and the ascospores are hyaline. The conidial state is in the genus *Penicillium*. Examine the bright yellow ascocarps of *Talaromyces flavus* (= *T. vermiculatus*) under the dissecting microscope. Mount several ascocarps in water and crush gently; examine the asci and ascospores (Fig. 106-107). Asci will be found only in young material. Also examine the conidial state, *Penicillium vermiculatum*.

Onygenales

In the Onygenales the ascocarp frequently consists of a loose arrangement of hyphae, and a distinct peridium is lacking. The anamorph is arthrosporic or aleuriosporic.

Gymnoascaceae — Cleistothecial wall consisting of a loose network of hyphae.

Gymnoascus — In *Gymnoascus* the peridial hyphae lack joints at the septa and the appendages are not well defined. The free ends of the hyphae are spine-like, with lateral branches. The ascospores are hyaline. Examine cultures of *Gymnoascus reessii* under the dissecting microscope. Mount several ascocarps in water and crush gently. Observe the peridial hyphae, asci, and ascospores (Fig. 112).

Microascales

This order includes ostiolate fungi having evanescent asci scattered at different levels in the ascocarp. The ascocarps are small and are formed on a well developed mycelium. The asci are usually eight-spored and deliquescent, freeing the ascospores inside the ascocarp. Mature ascospores are often pushed out of the ostiole in a cirrhus. These fungi are often found in soil and in association with wood. The order contains several serious plant pathogens.

Ophiostomataceae — Ascocarps very small, with very long ostiolar neck, usually partly immersed in the substrate.

Ceratocystis — The ascocarps are dark, very small, with a long ostiolar neck. The ascospores are one-celled and hyaline. Examine specimens of wood showing "blue-stain;" this is caused by several species, such as *Ceratocystis minor* and *C. pilifera.* Examine cultures of *C. fimbriata* (cause of black rot of sweet potato) under the dissecting microscope. Mount some ascocarps in water and crush gently. Look at the hat-shaped ascospores. Also look for phialides of the conidial state; these produce endoconidia of two types, cylindrical and barrel-shaped. Both types of conidia are hyaline and formed in chains. (Fig. 114-118). Examine the demonstration of elm bark beetles, beetle galleries, and infected elm wood associated with *C. ulmi*, cause of Dutch elm disease. Examine a culture of *Pesotum (Graphium) ulmi*, the conidial state of C. *ulmi*, under the dissecting microscope. Mount some synnemata in water and examine under the microscope (Fig. 119).

REFERENCES

Benjamin, C.R. 1955. Ascocarps of *Aspergillus* and *Penicillium. Mycologia* 47: 669-687.

Benjamin, R.K. 1956. A new genus of Gymnoascaceae with a review of other genera. *Aliso* 3: 301-328.

Griffin, H.D. 1966. The genus *Ceratocystis* in Ontario. *Canad. J. Bot.* 46: 689-718.

Horie, Y. 1980. Ascospore ornamentation and its application to the taxonomic re-evaluation in Emericella. *Trans. Mycol. Soc. Japan* 21: 483-493.

Hunt, John. 1956. Taxonomy of the genus *Ceratocystis. Lloydia* 129: 1-59.

Kuehn, H.H. 1958, 1959. A preliminary survey of the Gymnoascaceae. I, II. *Mycologia* 50: 417-439; 51: 665-692.

Olchowecki, A., and J. Reid. 1974. Taxonomy of the genus *Ceratocystis* in Manitoba. Canad. J. Bot. 52: 1675-1711.

Orr, G.F., and H.H. Kuehn. 1963. The genus *Ctenomyces* Eidam. *Mycopath. Mycol. Appl.* 21: 321-333.

Orr, G.F., H.H. Kuehn and O.A. Plunkett. 1963. The genus *Gymnoascus* Baranetzky. *Mycopath. Mycol. Appl.* 21: 1-18.

Orr, G.F., H.H. Kuehn, and O.A. Plunkett. 1963. The genus *Myxotrichum* Kunze. *Canad. J. Bot.* 41: 1457-1480.

Padhye, A.A., and J.W. Carmichael. 1971. The genus *Arthroderma* Berkeley. *Canad. J. Bot.* 49: 1525-1540.

Pitt, J.I. 1979. *The Genus Penicillium and its Teleomorphic States Eupenicillium* and *Talaromyces.* Academic Press, London.

Ramirez, C. 1982. *Manual and Atlas of the Penicillia.* Elsevier Biomedical Press, Amsterdam.

Raper, K.B., and C. Thom. 1949. *A Manual of the Penicillia.* Williams & Wilkins Co., Baltimore.

Raper, K.B., and D.I. Fennell. 1965. *The Genus Aspergillus.* Williams & Wilkins Co., Baltimore.

Rosinski, Martin A. 1961. Development of the ascocarp of *Ceratocystis ulmi. Amer. J. Bot.* 48: 285-293.

Scott, D.B., and A.C. Stolk. 1967. Studies on the genus *Eupenicillium* Ludwig. II. Perfect States of some *Penicillia. Antonie van Leeuwenhoek* 33: 297-314.

Stockdale, P.M. 1961. *Nannizzia incurvata* gen. nov., sp. nov., a perfect state of *Microsporum gypseum* (Bodin) Guiart et Grigorakis. *Sabouraudia* 1:41-48.

Stolk, A.C. 1968. Studies on the genus *Eupenicillium* Ludwig. III. Four new species of *Eupenicillium. Antonie van Leeuwenhoek* 34: 37-53.

Stolk, A.C., and R.A. Samson. 1972. The genus *Talaromyces.* Studies on *Talaromyces* and related genera II. *Stud. Mycol.* 2: 1-65.

Stolk, A.C., and R.A. Samson. 1983. The ascomycete genus *Eupenicillium* and related *Penicillium* anamorphs, *Stud. Mycol.* 23: 1-149.

Stolk, A.C., and D.B. Scott. 1967. Studies on the genus *Eupenicillium* Ludwig. I. Taxonomy and nomenclature of *Penicillia* in relation to their sclerotioid ascocarpic states. *Persoonia* 4: 391-405.

Upadhyay, H.P. 1981. A *Monograph of Ceratocystis and Ceratocytiopsis.* Univ. Georgia Press, Athens.

Pyrenomycetidae

In the Pyrenomycetidae, or pyrenomycetes, the asci are borne in a layer or fasicle in the ascocarp. A few species form only a single ascus per ascocarp. In most pyrenomycetes the ascocarp is an ostiolate, flask-shaped perithecium, but in some species the ascocarp is completely enclosed. The internal tissues of the ascocarp comprise the centrum, and in ostiolate species the ostiole is lined with slender periphyses. The make-up of the centrum tissues is variable and modern pyrenomycete taxonomy utilizes the different types of centrum tissues as the basis for distinguishing orders. Pyrenomycetous asci are variable in shape and usually contain eight ascospores. In most genera the asci are persistent, with a characteristic apical apparatus through which the ascospores are forcibly discharged.

The pyrenomycetes occur in a variety of habitats, but are especially common on dead wood, herbaceous stems, and in soils. Many species are important pathogens of economic crops.

Erysiphales

Members of the Erysiphales have completely enclosed ascocarps that contain one or several asci. In most genera the ascocarps are dark- brown and they bear a ring of characteristic appendages around the base, but in one tropical genus the ascocarp wall is thin and hyaline and appendages are lacking.

All are obligate parasites of higher plants, forming a superficial, hyaline mycelium that produces haustoria. In one tropical species the mycelium also bears erect, hyaline setae. In most species, young ascocarps are globose and hyaline when young, becoming yellow, then dark brown to black at maturity. Mature ascocarps are round in face view, and slightly flattened in side view. The ascocarp wall is composed of thick-walled pseudoparenchyma cells. Asci are produced singly or in a layer and are surrounded by a layer of thin-walled pseudoparenchyma tissue when young. The asci are broadly oval, with a thickened apex. Ascospores in the Erysiphales are all one-celled, hyaline, and oval. Conidial production in the Erysiphales is profuse, with powdery masses being formed on the host surface and giving rise to the common name "powdery mildew." Individual conidiophores are hyaline, erect, and simple, and produce conidia either singly or in a chain. The conidia are one-celled and hyaline, and vary in shape from oval to angular. The conidial state usually occurs early in the growing season, with the ascocarps forming later. Genera in the Erysiphales are separated on the number of asci per ascocarp and on the nature of the appendages if present. There is a single family, the **Erysiphaceae**.

Erysiphe graminis — The genus *Erysiphe* has ascocarps containing many asci and mycelioid appendages. *Erysiphe graminis* is common on grasses, especially cereal grains. Examine prepared slides of sections through immature ascocarps and note the thick-walled pseudoparenchymatous ascocarp wall, the layer of thin-walled pseudoparenchyma inside the wall, and the young asci (Fig. 120-121, and 126). Examine other slides of the conidial state, *Oidium*, and note how the conidia form in chains (Fig. 122-123). Mount several ascocarps from dried material in a drop of water. Observe the appendages, then crush gently and look for asci, and ascospores, if present.

Microsphaera — In this genus the ascocarp contains many asci and the tips of the appendages are dichotomously branched. Mount several ascocarps in a drop of water and observe the appendages (Fig. 131).

Phyllactinia — In this genus there are many asci and the appendages are stiff, with a bulbous base. Mount several ascocarps in a drop of water and observe the appendages (Fig. 130). If material of the conidial state is available, mount some conidiophores in water and note the single conidium. The conidial state belongs in the genus *Ovulariopsis*.

Uncinula — The ascocarps in *Uncinula* contain many asci and the appendages are rigid, with an uncinate or recurved tip. Mount several ascocarps in a drop of water and observe the appendages. Crush the ascocarps gently and look for asci and ascospores (Fig. 124-125, 128-129).

Podosphaera —Ascocarps of this genus contain a single ascus and have dichotomously branched appendages. Examine prepared slides showing sections through ascocarps containing a single large ascus. Mount several ascocarps from dried material in a drop of water and observe the appendages (Fig. 127).

Oidium — If available, mount fresh material of the conidial state, *Oidium*, which has chains of conidia on simple conidiophores. On the basis of the conidial state alone it is often not possible to distinguish the perfect genera.

REFERENCES

Blumer, S. 1967. *Echte Mehltaupilze (Erysiphaceae)*. Gustav Fischer Verlag, Jena.

Boesewinkel, H.J. 1977. Identification of Erysiphaceae by conidial characteristics. *Rev. Mycol.* 41: 493-507.

Boesewinkel, H.J. 1980. The morphology of the imperfect states of powdery mildews (Erysiphaceae). *Bot. Rev.* 46: 167-224.

Bracker, C.E. 1968. Ultrastructure of the haustorial apparatus of *Erysiphe graminis* and its relationship to the epidermal cell of barley. *Phytopathology* 58: 12-30.

Braun, U. 1980. Morphological studies in the genus *Oidium*. *Flora* 170: 77-90.

Braun, U. 1982. Morphological studies in the genus *Oidium*. (II). *Zbl. Mikrobiol.* 137: 138-152.

Braun, U. 1982. Morphological studies in the genus *Oidium*. (III). *Zbl. Mikrobiol.* 137: 314-324.

Colson, B. 1938. The cytology and development of *Phyllactinia corylea* Lev. *Ann. Bot. n.s.* 2: 381-402.

Hanlin, R.T., and O. Tortolero. 1984. An unusual tropical powdery mildew. *Mycologia* 76: 439-442.

Hirata, K. 1966. *Host range and geographical distribution of the powdery mildews.* Pub. by College of Agri., Niigata Univ., Japan.

Hodges, C.S., Jr. 1985. Hawaiian forest fungi VI. A new species of *Brasiliomyces* on *Sapindus ohauensis*. *Mycologia* 77: 977-981.

Mitchell, N.L., and W.E. McKeen. 1970. Light and electron microscope studies on the conidium and germ tube of *Sphaerotheca macularis*. *Canad. J. Microbiol.* 16: 273-280.

Pady, S.M., C.L. Kramer, and R. Clary. 1969. Sporulation in some species of *Erysiphe*. *Phytopathology* 59: 844-848.

Pady, S.M., and J. Subbayya. 1970. Spore release in *Uncinula*. *Phytopathology* 60: 1702-1703.

Parmalee, J.A. 1977. The fungi of Ontario. II. Erysiphaceae (mildews). *Canad. J. Bot.* 55: 1940-1983.

Salmon, E.S. 1900. A monograph of the Erysiphaceae. *Mem. Torrey Bot. Club* 9: 1-292.

Spencer, D.M. (ed.). *The Powdery Mildews*. Academic Press, New York.

Yarwood, C.E. 1957. Powdery mildews. *Bot. Rev.* 23: 235-300.

Zheng, R-Y. 1985. Genera of the Erysiphaceae. *Mycotaxon* 22: 209-263.

FIG. 120. Ascocarp of *Erysiphe graminis* with mycelioid appendages. x 140. FIG. 121. Section through host leaf (HL) bearing ascocarp of *Erysiphe graminis* with three young asci (AS) surrounded by thin-walled pseudoparenchyma (TWP) and ascocarp wall (AW). MYC = mycelium. x 440. FIG 122. Two young conidia of *Oidium* on conidiophore. x 200. FIG. 123. Mature conidium of *Oidium*. x 200. FIG. 124. Ascocarp of *Uncinula macrospora* with appendages. x 160. FIG. 125. Close-up of appendages of *Uncinula macrospora* x 520. FIG. 126. Section through ascocarp of *Erysiphe graminis* with asci containing young ascospores. x 400. FIG. 127. Section through ascocarp of *Podosphaera* showing single immature ascus. x 680. FIG. 128. Ascus of *Uncinula macrospora* containing two young ascospores. x 600. FIG. 129. Mature ascospore of *Uncinula macrospora*. x 800.

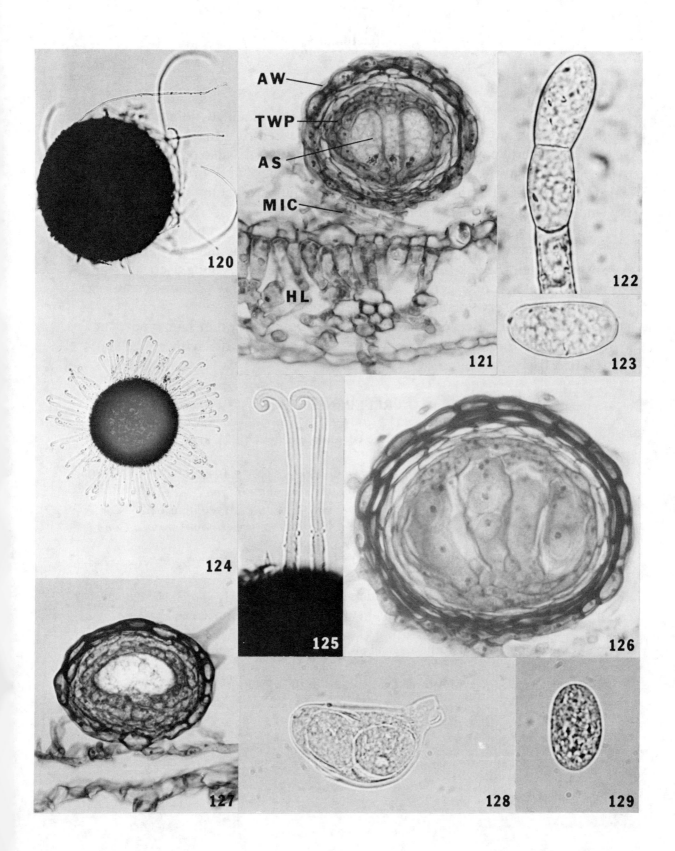

Chaetomiales

Members of this order have dark brown to black, ostiolate ascocarps that are covered by characteristic hairs. The centrum in young ascocarps contains lateral paraphyses, with hymenial paraphyses also reported in some species. The asci are evanescent and the one-celled, dark brown ascospores are extruded in a cirrhus. Most species are cellulose decomposers. There is a single family, the **Chaetomiaceae.**

Chaetomium globosum — This is one of the most common species and has wavy hairs. Examine cultures under the dissecting microscope and observe the hairs. Look for cirrhi of ascospores. Mount young ascocarps in a drop of water, crush gently, and examine for asci and young ascospores. Mount and observe mature ascospores (Fig. 132-134).

Mount and observe species of *Chaetomium* with other hair types, such as *C. funicolum* with branched hairs and *C. cochliodes* with coiled hairs (Plate VI).

REFERENCES

Ames, L.M. 1961. A Monograph of the Chaetomiaceae. *U.S. Army Res. & Dev.* Ser. No. 2:1-125.

Cooke, J.C. 1969. Morphology of *Chaetomium erraticum. Amer. J. Bot.* 56:335-340.

Cooke, J.C. 1969. Morphology of *Chaetomium funicolum. Mycologia* 61:1060-1065.

Cooke, J.C. 1970. Morphology of *Chaetomium trilaterale. Mycologia* 62:282-288.

Corlett, M. 1966. Perithecium development in *Chaetomium trigonosporum. Canad. J. Bot.* 44:155-162.

Seth, H.K. 1970 (1972). A monograph of the genus *Chaetomium. Beih. Nova Hedwigia* 37:1-133.

Whiteside, W.C. 1957. Perithecial initials of *Chaetomium. Mycologia* 49:420-425.

Whiteside, W.C. 1961. Morphological studies in the Chaetomiaceae. I. *Mycologia* 53:512-523.

Whiteside, W.C. 1962. Morphological studies in the Chaetomiaceae. II. *Mycologia* 54:152-159.

Whiteside, W.C. 1962. Morphological studies in the Chaetomiaceae. III. *Mycologia* 54:611-620.

Phyllachorales

Members of the Phyllachorales are mostly parasitic on higher plants, on which they cause diseases, especially on the leaves. The ascomata are often immersed in host tissues. They are all ostiolate and usually obpyriform, with filamentous paraphyses and hyaline, one-celled ascospores. Centrum pseudoparenchyma is lacking. True stromata are lacking, but the perithecia are often associated with a clypeus, a thin stromatic layer that surrounds the ostioles. Anamorphs are present and some species regularly form spermatia. There is a single family, the **Polystigmataceae (Phyllachoraceae)**.

Glomerella — The perithecia in this genus are not in a stroma, but are often immersed in host tissues. The ascospores are one-celled and hyaline. Mount several perithecia from culture and look for asci and ascospores. The ascospores are one-celled and curved (Fig. 156-157). Also look for conidia of the imperfect state, *Colletotrichum*; they are hyaline and straight.

Phyllachora graminis — This is a common parasite of grasses. Examine dried material under the dissecting microscope and note the characteristic lesion. Examine prepared slides showing sections of host leaf bearing ascocarps and look for paraphyses, asci, and ascospores. The ascospores are one-celled and hyaline (Fig. 138).

REFERENCES

Boyd, E.S. 1934. A developmental study of a new species of Ophiodothella. Mycologia 26: 456-468.

Chilton, S.J.P., G.B. Lucas, and C.W. Edgerton. 1945. Genetics of *Glomerella*. III. Crosses with a conidial strain. *Amer. J. Bot.* 32: 549-554.

Chilton, S.J.P., and H.E. Wheeler. 1949. Genetics of *Glomerella*. VI. Linkage. *Amer. J. Bot.* 36: 270-273.

Chilton, S.J.P. and H.E. Wheeler. 1949. Genetics of *Glomerella* VIII. Mutation and segregation in plus cultures. *Amer. J. Bot.* 36: 717-721.

Edgerton, C.W., S.J.P. Chilton, and G.B. Lucas. 1945. Genetics of *Glomerella*. II. Fertilization between strains. Amer. J. Bot. 32: 115-118.

Lucas, G.B. 1946. Genetics of *Glomerella*. IV. Nuclear phenomena in the ascus. Amer. J. Bot. 33: 802-806.

Lucas, G.B., S.J.P. Chilton, and C.W. Edgerton. 1944. Genetics of *Glomerella*. I. Studies on the behavior of certain strains. *Amer. J. Bot.* 31: 233-239.

McGahen, J.W., and H.E. Wheeler. 1951. Genetics of *Glomerella*. IX. Perithecial development and plasmogamy. Amer. J. Bot. 38: 610-617.

Miller, J.H. 1951. Studies in the Phyllachoraceae. I. *Phyllachora ambrosiae* (Berk. & Curt.) Sacc. *Amer. J. Bot.* 38: 830-834.

Miller, J.H. 1954. Studies in the Phyllachoraceae. *Phyllachora*. II. *Phyllachora lespedezae*. *Amer. J. Bot.* 41: 825-828.

Orton, C.R. 1944. Graminicolous species of *Phyllachora* in North America. *Mycologia* 36: 18-53.

Pady, S.M., and C.L. Kramer. 1971. Spore discharge in *Glomerella*. *Trans. Brit. Mycol. Soc.* 56: 81-87.

Parbery, D.G. 1963. Studies on graminicolous species of *Phyllachora* Fckl. I. Ascospores — their liberation and germination. *Aust. J. Bot.* 11: 117-130.

Parbery, D.G. 1963. Studies on graminicolous species of *Phyllachora* Fckl. II. Invasion of the host and development of the fungus. *Aust. J. Bot.* 11: 131-140.

Parbery, D.G. 1967. Studies on graminicolous species of *Phyllachora* Nke. in Fckl. V. A taxonomic monograph. *Aust. J. Bot.* 15: 271-375.

Parbery, D.G., and R.F.N. Langdon. 1963. Studies on graminicolous species of *Phyllachora* Fckl. III. The relationship of certain scolecospores to species of *Phyllachora*. *Aust. J. Bot.* 11: 141-151.

Parbery, D.G., and R.F.N. Langdon. 1964. Studies on graminicolous species of *Phyllachora* Fckl. IV. Evaluation of the criteria of species. *Aust. J. Bot.* 12: 265-281.

Wheeler, H.E. 1956. Sexual versus asexual reproduction in *Glomerella*. *Mycologia* 48: 349-353.

Wheeler, H.E. 1956. A genetic basis for the occurrence of minus mutants. *Amer. J. Bot.* 37: 304-312.

Wheeler, H.E., L.S. Olive, C.T. Ernest and C.W. Edgerton. 1948. Genetics of *Glomerella*. V. Crozier and ascus development. *Amer. J. Bot.* 35: 722-728.

FIG. 130. Appendage with bulbous base of *Phyllactinia*. × 250. FIG. 131. Branched appendage tip of *Microsphaera*. × 400. FIG. 132-134. *Chaetomium globosum*. FIG. 132. Ascocarp with wavy hairs and attachment hyphae at base. × 100. FIG. 133. Young ascus with immature ascospores. × 680. FIG. 134. Mature ascospores. × 680. FIG. 135-137. *Melanospora zamiae*. FIG. 135. Mature ascocarp; ascospores inside are visible through transparent wall. × 160. FIG. 136. Young ascus (arrow) with immature ascospores. × 1000. FIG. 137. Mature ascospores. × 680. FIG. 138. Section through ascocarp of *Phyllachora graminis* in host leaf (HL) showing ascocarp wall (AW), ostiole (O), ostiolar neck (ON), asci (AS) with ascospores, and paraphyses (PA). × 400. FIG. 139-141. *Thielavia terricola*. FIG. 139. Mature ascocarp. × 225. FIG. 140. Young ascus with immature ascospores. × 400. FIG. 141. Mature ascospores. × 800.

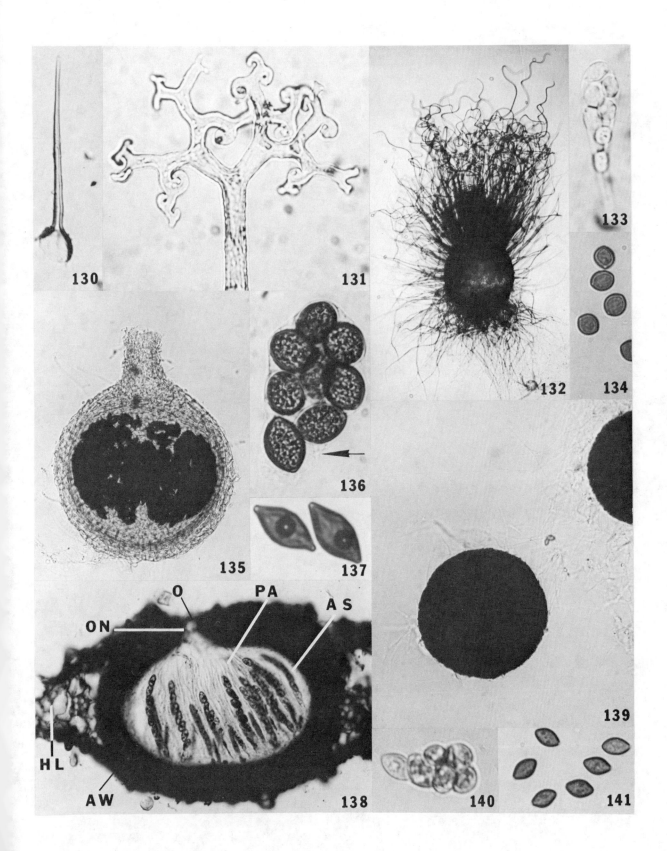

Sordariales

In the Sordariales the ascoma is a non-stromatic perithecium with non-carbonaceous, pseudoparenchymatous walls, with a centrum composed of pseudoparenchyma or a combination of pseudoparenchyma and paraphyses. In some species the wall is thin and translucent. Paraphyses are usually filamentous when young, becoming vesiculose with age. Asci may be prototunicate or unitunicate; when unitunicate, a non-amyloid apical ring is typically present. In most species the ascospores are unicellular and dark brown, with a germ pore or a germ slit, but in some genera the ascospores are two-celled, at least when young. Anamorphs are phialidic or blastogenous. Sordariaceous fungi are saprobic, often coprophilous, but they also occur in soils and seeds. Two families are considered here, the **Melanosporaceae** and the **Sordariaceae.**

Melanosporaceae — Centrum pseudoparenchymatous, paraphyses lacking, ascospores with germ pores, asci deliquescent.

Melanospora — This genus has brown, transparent perithecia with evanescent asci and one-celled, brown ascospores. It usually grows in association with another fungus; it is not parasitic but derives some essential nutrients from it. Mount some perithecia of *Melanospora zamiae* in water and examine the asci and ascospores. Asci will be foung only in young material (Plate VI, Fig. 135-137).

Thielavia terricola — This is a common soil fungus with dark brown, completely enclosed ascocarps and evanescent asci. The ascospores are dark brown and one-celled (Fig. 139-141). Mount young ascocarps and look for asci and ascospores, then mount mature ascocarps and examine the ascospores. Because of the enclosed asocarp this genus has traditionally been placed in the plectomycetes, but recently some workers have moved it to the pyrenomycetes because of the ascocarp morphology and the relatively large ascospores which have a germ pore.

Sordariaceae — Centrum with pseudoparenchyma inside wall, paraphyses filamentous when young, vesiculose at maturity, asci persistent, ascospores with a germ pore.

Sordaria — The one-celled, dark brown ascospores are smooth and are surrounded by a gelatinous sheath. Mount small portions of cellulose squares from young cultures of *Sordaria fimicola* and look for coils, which represent the beginning of ascocarp formation. Mount and crush some mature ascocarps. Examine the asci and note the characteristic ring at the apex. Examine the ascospores and note the gelatinous sheath and the germ pore in the end of the spore. If available, examine sections through mature ascocarps and note the ascocarp wall, arrangement of asci, and the periphyses lining the ostiolar neck. Mount some perithecia from wild type (dark brown) X gray ascospore crosses and examine the arrangement of ascospores in the asci, which is indicative of the presence or absence of meitoic crossing over (Plate VI, Fig. 142-150).

REFERENCES

Arx., J.A. von. 1975. On Thielavia and some similar genera of Ascomycetes. *Stud. Mycology* 8: 1-29.

Booth, C. 1961. Studies of Pyrenomycetes: VI. *Thielavia,* with notes on some allied genera. *C.M.I. Mycol. Papers* 83: 1-15.

Cain, R.F. 1950. Studies of coprophilous Ascomycetes. I..Gelasinospora. Canad. J. Res. C. 28: 566-576.

Cain, R.F., and J.W. Groves. 1948. Notes on seed-borne fungi. VI. *Sordaria. Canad. J. Res. C.* 326: 486-495.

Cannon, P.F., and D.L. Hawksworth. 1982. A re-evaluation of *Melanospora* Corda and similar Pyrenomycetes, with a revision of British species. *Bot. J. Linn. Soc.* 84: 115-160.

Doguet, G. 1955. Le genre *Melanospora:* biologie, morphologie, developpment, systematique. *Botaniste* 39: 1-313.

Emmons, C.W. 1932. The development of the ascocarp in two species of *Thielavia. Bull. Torrey Bot. Club* 59: 415-422.

Jensen, J.D. 1982. The development of *Gelasinospora reticulospora. Mycologia* 74: 724-737.

Kowalski, D.T. 1965. The development and cytology of *Melanospora tiffanii. Mycologia* 57: 279-290.

Lu, B.C. 1967. The course of meiosis and centriole behavior during the ascus development of the ascomycete *Gelasinospora calospora. Chromosoma* 22: 210-226.

Lundqvist, N. 1972. Nordic Sordariaceaes. lat. *Symb. Bot. Upsal.* 20 (1): 1-374.

Mirza, J.H., and R.F. Cain. 1969. Revision of the genus *Podospora.* Canad. J. Bot. 47: 1999-2048.

Nelson, A.C., and M.P. Backus. 1968. Ascocarp development in two homothallic *Neurosporas.* Mycologia 60: 16-28.

Olive, L.S. 1956. Genetics of *Sordaria fimicola.* I. Ascospore color mutants. *Amer. J. Bot..* 43: 97-107.

Reeves, F.B. 1971. The structure of the ascus apex in *Sordaria fimicola. Mycologia* 63: 204-212.

Sanni, M.O. 1982. Perithecium development in *Gelasinospora tetrasperma. Mycologia* 74: 320-324.

Uecker, F.A. 1976. Development and cytology of *Sordaria humana. Mycologia* 68:

FIG. 142-150. *Sordaria fimicola*. FIG. 142. Coiled ascocarp initial. × 800. FIG. 143. Mature ascocarp. × 140. FIG. 144. Section through mature ascocarp with ascocarp wall (AW), ostiolar neck (ON), ostiole (O), periphyses (PE), and asci (AS) with ascospores. × 185. FIG. 145. Mature ascus with eight ascospores. × 615. FIG. 146. Mature ascus containing 4:4 arrangement of light and dark ascospores, indicating absence of meiotic crossing-over. × 400. FIG. 147. Mature asci with 2:4:2 and 2:2:2:2:2 arrangements of light and dark ascospores, indicating meiotic crossing-over has occurred. × 400. FIG. 148. Optical section through ring at apex of mature ascus. × 1550. FIG. 149. Germ pore in end of mature ascospore. × 1100. FIG. 150. Mature ascospore with sheath stained in cotton blue. × 1000. FIG. 151-153. *Hypoxylon* sp. FIG. 151. Germ slit in mature ascospore. × 1000. FIG. 152. Side view of mature ascospore with two oil droplets. × 1000. FIG. 153. Mature ascus with eight ascospores. × 320. FIG. 154-155. *Xylaria hypoxylon*. FIG. 154. Cross section through mature stroma with white interior tissue, black outer layer, and perithecia around periphery. × 16. FIG. 155. Section through mature perithecium in stroma (SR) showing ascocarp wall (AW), asci (AS) with ascospores, and ostiole (O). × 160.

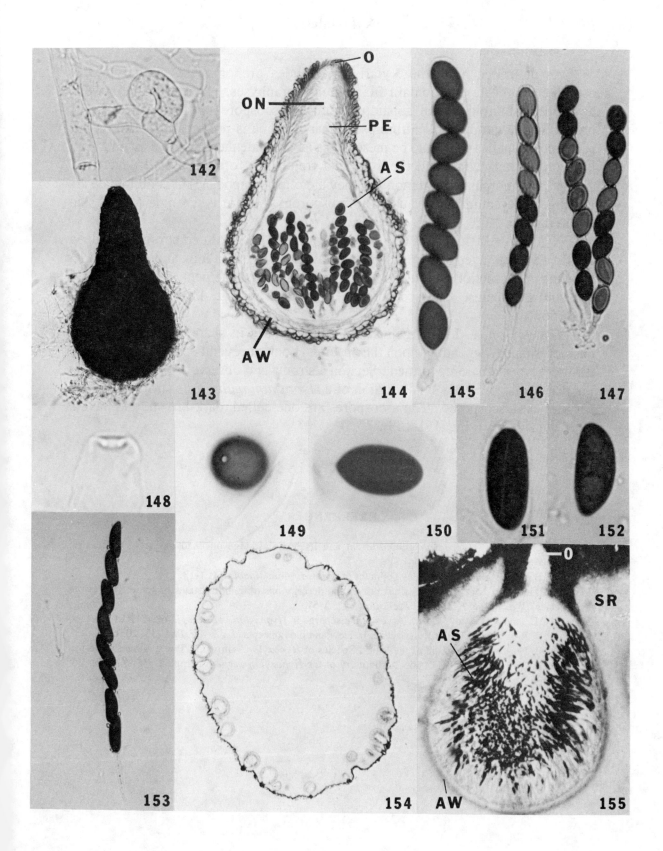

Xylariales

As delimited here, the Xylariales includes prenomycetes with typically ostiolate perithecia that contain filamentous paraphyses. The asci have a well-developed, amyloid, apical apparatus, and the ascospores are one-celled, dark-brown, with a germ slit. Centrum pseudoparenchyma is presumably lacking, but some recent studies show it to be present as a layer inside the wall. The ascoma wall is often carbonaceous, and the ascomata frequently form in a well-developed stroma. Anamorphs are common and they are often sympodial, but conidiomata are lacking. The conidia are usually amerosporous. There is a single family, the **Xylariaceae**.

Xylaria — Members of this genus form erect, dark-colored stromata on dead wood. Examine dried specimens of *Xylaria hypoxylon*, then look at prepared slides showing cross sections through a stroma. Note the arrangement of the perithecia, asci, and ascospores. The ascospores are one-celled and dark brown (Plate VI, Fig. 154-155).

Hypoxylon — In this genus the stromata vary in shape from flattened to cushion-shaped, depending upon the species. Look at several species and note the different types of stroma formed. Section a stroma and observe the arrangement of the perithecia. Mount a small portion of a *Hypoxylon multiforme* stroma and look for asci and ascospores. The ascospores are one-celled, dark-brown, and have a germ slit (Plate VI, Fig. 151-153).

REFERENCES

Barnett, H.L. 1957. *Hypoxylon punctulatum* and its conidial stage on dead oak trees in culture. *Mycologia* 49: 588-595.

Brown, H.B. 1913. Studies in the development of *Xylaria*. *Ann. Mycol.* 11: 1-13.

Jong, S.C., and J.D. Rogers. 1972. Illustrations and descriptions of conidial states of some *Hypoxylon* species. *Wash. Agr. Expt. Sta. Tech. Bull.* 71: 1-51.

Kramer, C.L., and S.M. Pady. 1970. Ascospore discharge in *Hypoxylon*. *Mycologia* 62: 1170-1186.

Martin, P. 1967. Studies in the Xylariaceae: I. New and old concepts. *J. So. Afr. Bot.* 33: 205-240.

Miller, J.H. 1961. A Monograph of the World Species of *Hypoxylon*. Univ. Ga. Press, Athens. 158 pp.

Petrini, L.E., and J.D. Rogers. 1986. A summary of the *Hypoxylon serpens* complex. *Mycotaxon* 26: 401-436.

Rogers, J.D. 1965. *Hypoloxon fuscum*. I. Cytology of the ascus. *Mycologia* 57: 789-803.

Rogers, J.D. 1969. *Hypoxylon rubiginosum*: cytology of the ascus and surface morphology of the ascospore. *Mycopath. Mycol. Appl.* 38: 215-223.

Rogers, J.D. 1969. *Xylaria polymorpha*. I. Cytology of a form with small stromata from Minnesota. *Canad. J. Bot.* 47: 1315-1317.

Rogers, J.D. *Hypoxylon cohaerens*: cytology of the ascus. *Mycopath. Mycol. Appl.* 48: 161-165.

Rogers, J.D. *Xylaria cubensis* and its anamorph *Xylocoremium flabelliforme*, *Xylaria allantoidea*, and *Xylaria poitei* in continental United States. *Mycologia* 76: 912-923.

Rogers, J.D. 1986. Provisional keys to *Xylaria* in continental United States. *Mycotaxon* 26: 85-97.

Rogers, J.D. and B.E. Callan. 1986. *Xylaria poitei*: stromata, cultural description, and structure of conidia and ascospores. *Mycotaxon* 26: 287-296.

Rogers, J.D., and B.E. Callan. 1986. *Xylaria polymorpha* and its allies in continental United States. Mycologia 78: 391-400.

Thite, A.N. 1977. Ascospore formation in *Xylaria apiculata*. *Trans. Brit. Mycol. Soc.* 69: 148-150.

Whalley, A.J.S. and G.N. Greenhalgh. 1973. Numerical taxonomy of Hypoxylon I. Comparison of classifications of the cultural and the perfect states. *Trans. Brit. Mycol. Soc.* 61: 435-454.

PLATE VI. TOP: Left, *Chaetomium cochliodes.* × 100. Right, *Chaetomium funicolum.* × 100. Center, *Melanospora zamiae.* × 100. Lower right, *Sordaria fimicola.* × 100.

BOTTOM: Upper left, Stroma of *Hypocrea rufa.* × 10. Upper center, Stroma of *Hypoxylon multiforme.* × 10. Lower center, Perithecia of *Nectria cinnabarina* with sporodochia of the conidial state, *Tubercularia vulgaris* on bottom and left of branch. × 10. Right, Stromata of *Xylaria hypoxylon.* × 1.

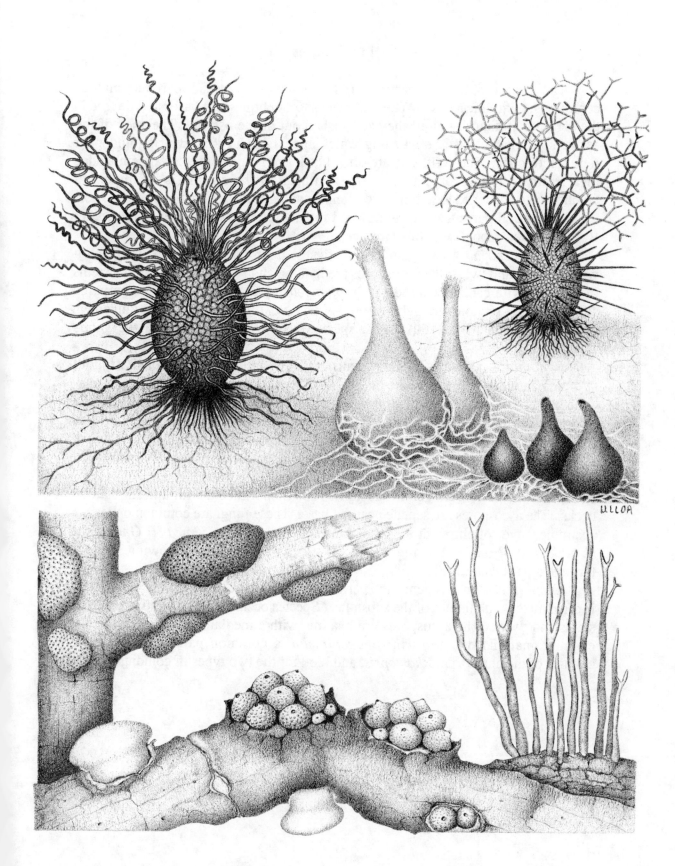

Diaporthales

The Diaporthales are perithecial fungi having a pseudoparenchymatous centrum and unitunicate asci. The asci have short deliquescent bases and lie free in the centrum at maturity, until they are pushed out of the ostiole. The apex of the ascus has a characteristic thickening which appears as a refractive ring under the light microscope. The perithecia are often black and in many species are embedded in a stroma. The arrangement of the peritheca in the stroma is an important taxonomic character. These fungi are often found on leaves and woody tissues, both as saprobes and parasites. All are placed in a single family, the **Diaporthaceae**. Barr, however, recognizes four families.

Cryphonectria — This genus forms fleshy, light-colored valsoid stromata on woody stems and twigs; the ascospores are hyaline and two celled. Examine stems of chestnut with cankers caused by *Cryphonectria parasitica*. Examine prepared slides showing sections through a stroma of *C. parasitica* and note the arrangement of the perithecia (Fig. 160). This species was formerly placed in the genus *Endothia*.

Endothia — Species in this genus form fleshy, light-colored, diatrypoid stromata on woody stems., twigs and roots. The ascospores are hyaline and one-celled. Examine oak twigs bearing conidial and perithecial stromata of *Endothia gyrosa* under the dissecting microscope. Mount several peritheca in water and look for asci and ascospores (Fig. 159). Examine a culture of *Endothiella,* the conidial state of *Endothia gyrosa*. Mount some pycnidia in water and observe under the microscope.

Stegophora — This genus has non-stromatic perithecia with upright ostiolar necks. The ascospores are septate near the base. These fungi are common on leaves. Examine leaves of elm with leaf spot caused by *Stegophora ulmea [= Gnomonia (Dothidella) ulmea]*. Examine prepared slides of this species showing sections through a perithecium (FIg. 158).

Diaporthe — In this genus the perithecia are surrounded by a dark stroma that is often mixed with tissues of the substrate. Species occur on woody and herbaceous stems and twigs. The ascospores are hyaline, with a median septum. If available, examine material of *Diaporthe phaseolorum,* a common pathogen. Examine cultures of the anamorph, *Phomopsis*, and look for the two types of conidia.

REFERENCES

Anagnostakis, S.L. 1979. Sexual reproduction of *Endothia parasitica* in the laboratory. Mycologia 61: 213-215.

Anagnostakis, S.L. 1982. The origin of ascogenous nuclei in *Endothia parasitica*. Genetics 100: 413-416.

Anagnostakis, S.L. 1982. An improved defined medium for growth of *Endothia parasitica*. *Mycologia* 74: 826-830.

Anagnostakis, S.L. 1982. Biological control of chestnut blight. *Science* 215: 466-471.

Barr, M.E. 1978. The Diaporthales in North America. *Mycol. Mem.* 7: 1-232.

Huang, L.H., and E.S. Luttrell. 1982. Development of the perithecium in Gnomonia comari (Diaporthaceae). *Amer. J. Bot.* 69: 421-431.

Jensen, J.D. 1983. The development of *Diaporthe phaseolorum* variety *sojae* in culture. *Mycologia* 75: 1074-1091.

Kobayashi, T. 1970. Taxonomic studies of Japanese Diaporthaceae with special reference to their life-histories. *Bull. Govt. For. Exp. Sta.* 226: 1-242.

Morgan-Jones, J.F. 1953. Morpho-cytological studies of the genus *Gnomonia*. I. *Svensk Bot. Tid.* 47: 284-308.

Morgan-Jones, J.F. 1953. Morpho-cytological studies of the genus *Gnomonia*. II. *Svensk Bot. Tid.* 52: 363-372.

Morgan-Jones, J.F. 1953. Morpho-cytological studies of the genus *Gnomonia*. III. *Svensk Bot. Tid.* 53: 31-101.

Wehmeyer, L.E. 1933. The genus *Diaporthe* Nitschke and its segregates. *Univ. Michigan Stud. Sci. Ser.* 9: 1-349.

FIG. 156-157. *Glomerella cingulata*. FIG. 156. Ascus with ascospores. x 500. FIG. 157. Mature ascospores. x 500. FIG. 158. Section through perithecium of *Stegophora ulmea* on host leaf (HL) showing ascocarp wall (AW), ostiolar neck with ostiole, and centrum (CT) with young asci. x 240. FIG. 159. Ascus and ascospores of *Endothia gyrosa*. x 1000. FIG. 160. Section through perithecial stroma of *Cryphonectria parasitica* showing host tissue (HT), perithecia (PT) in stroma (SR), and ostiolar neck (ON). x 120. FIG. 161-162. *Hypomyces lactifluorum*. FIG. 161. Thickened ascus tip. x 1800. FIG. 162. Apiculate ascospore. x 1400. FIG. 163-166. *Nectria haematococca*. FIG. 163. Perithecium. x 160. FIG. 164. Ascus with ascospores. x 940. FIG. 165. Ascospore with longitudinal striations. x 2500. FIG. 166. Conidia of *Fusarium solani,* imperfect state of *Nectria haematococca.* x 1000. FIG. 167-168. *Claviceps purpurea*. FIG. 167. Section through portion of stalk (ST) bearing perithecial stroma (SR) containing perithecia (PT). x 50. FIG. 168. Close-up of perithecium in stroma (SR). Note asci around ostiole. x 200.

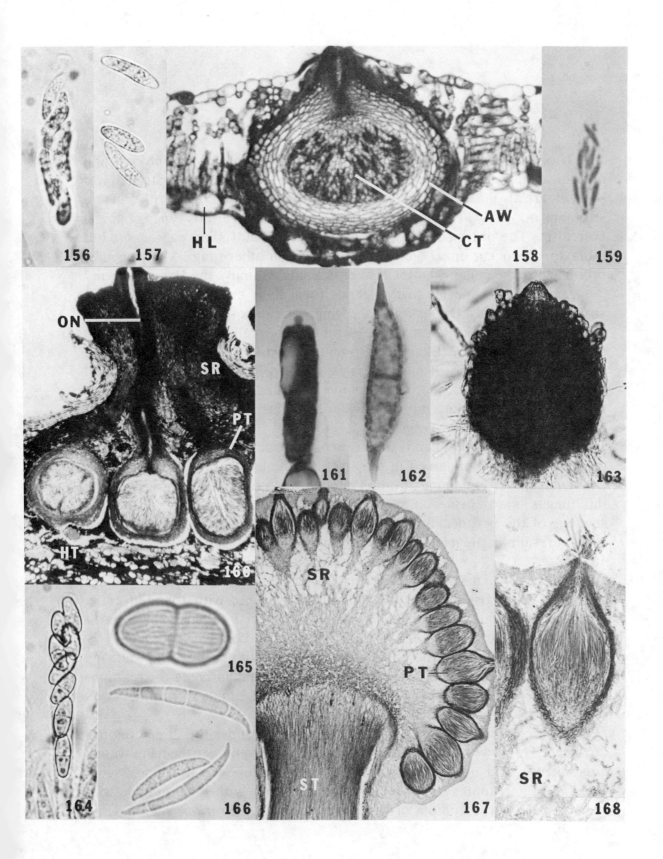

Hypocreales

The Hypocreales are perithecial fungi having apical paraphyses and unitunicate asci. Most species have brightly colored ascocarps; stromata, if present, are also usually brightly colored. The tissues of both perithecia and stromata are fleshy. The asci are persistent and have various types of apical structures. The order contains a number of important plant pathogens, but other species occur on wood and as parasites on other fungi. Many species produce profuse conidial states. There is a single family, the **Hypocreaceae**.

Hypocrea — Species in this genus form their perithecia immersed in stromata, with only the ostioles showing. The ascospores are two-celled with a marked constriction at the septum. The cells often separate into part-spores at maturity. Examine stromata of *Hypocrea* under the dissecting microscope (Plate VI). If material is available, mount some perithecia and look for asci and ascospores. Examine a culture of the imperfect state, *Trichoderma,* and make a mount in water and observe the conidiophores and conidia under the microscope.

Hypomyces — This genus is also stromatic, but usually only the base of the perithecium is immersed in the stroma. The ascospores are two-celled and apiculate. Examine mushrooms infected with *Hypomyces lactifluorum;* this fungus is a parasite on species of *Lactarius* and *Russula*. Make thin sections of the perithecial stroma and mount in water. Look for asci and ascospores under the microscope; note the structure of the ascus apex (Fig. 161-162).

Nectria — This genus contains both stromatic and non-stromatic species. The ascospores are two-celled and hyaline, with rounded ends. Examine twigs bearing both conidial and perithecial stromata of *Nectria cinnabarina* under the dissecting microscope. The conidial state is *Tubercularia vulgaris*. Examine the demonstration slides showing sections through conidial sporodochia and perithecial stromata and note the relationship between the perithecia and stroma. Examine cankers on hardwoods caused by *N. cinnabarina* infection (Plate VI).

Examine a culture of *Nectria haematoccocca* under the dissecting microscope. This is a non-stromatic species. Mount several perithecia in water, crush gently, and observe under the light microscope for asci and ascospores. The ascospores are two-celled and hyaline, with longitudinal striations. Also observe conidia of the imperfect state, *Fusarium solani* (Fig. 163-166).

REFERENCES

Booth, C. 1959. Studies of Pyrenomycetes: IV. *Nectria* (Part 1). *C.M.I. Mycol. Pap.* 73: 1-115.

Booth, C. 1960. Studies of Pyrenomycetes: V. Nomenclature of some *Fusaria* in relation to their nectrioid perithecial states. *CM.I. Mycol. Pap.* 74: 1-16.

Canham, S.C. 1969. Taxonomy and morphology of *Hypocrea citrina*. *Mycologia* 61: 315-331.

Doi, Y. 1969. Revision of the Hypocreales with cultural observations. IV. The genus *Hypocrea* and its allies in Japan (2) Enumeration of the species. *Bull. Nat. Sci. Mus. Tokyo* 15: 649-751.

Doi, Y. 1969. Revision of the Hypocreales with cultural observations. IV. The genus *Hypocrea* and its allies in Japan. *Bull. Nat. Sci. Mus. Tokyo* (1) General parts. 12: 694-724.

Ehrlich, J. 1934. The beech bark disease. A *Nectria* disease of *Fagus*, following *Cryptococcus fagi* (Baer). *Canad. J. Res.* 10: 593-692.

El-Ani, A.S. 1956. Ascus development and nuclear behavior in *Hypomyces solani* f. *cucurbitae*. *Amer. J. Bot.* 43: 769-778.

Hanlin, R.T. 1971. Morphology of *Nectria haematococca*. *Amer. J. Bot.* 58: 105-116.

Lohman, M.L., and A.J. Watson. 1943. Identity and host relations of *Nectria* species associated with diseases of hardwoods in the eastern states. *Lloydia* 6: 77-108.

Luttrell, E.S. 1944. Morphology of *Sphaerostilbe aurantiicola* (B. & Br.) Petch. *Bull. Torrey Bot. Club* 71: 599-619.

Rogerson, C.T. 1970. The hypocrealean fungi (Ascomycetes, Hypocreales). *Mycologia* 62: 865-910.

Rossman, A.Y. 1983. The phragmosporous species of *Nectria* and related genera. *CMI Mycol. Pap.* 150: 1-164.

Samuels, G.J. 1973. Perithecial development in *Hypomyces aurantius*. *Amer. J. Bot.* 60: 268-276.

Samuels, G.J. 1976. A revision of the fungi formerly classified as *Nectria* subgenus Hyphonectria. *Mem. N.Y. Bot. Gard.* 26(3): 1-126.

Seaver, F.J. 1909. The Hypocreales of North America — I. *Mycologia* 1: 41-76.

Seaver, F.J. 1909. The Hypocreales of North America — II. *Mycologia* 1: 177-207.

Seaver, F.J. 1910. The Hypocreales of North America — III. *Mycologia* 2: 48-92.

Smalley, E.B., and H.N. Hansen. 1957. The perfect stage of *Gliocladium roseum*. *Mycologia* 49: 529-533.

Sobers, E.K. 1969. *Calonectria floridana* sp. nov., the perfect stage of *Cylindrocladium floridanum*. *Phytopathology* 59: 364-366.

Clavicipitales

The Clavicipitales are perithecial fungi having lateral paraphyses and unitunicate asci. The perithecia are produced in a well developed stroma which is usually light colored. The asci are long and cylindrical, with a thickened apex, and they contain eight ascospores. The ascospores are filiform, hyaline, and septate; the cells often break apart in mounting them for study. Because of the light-colored stromata in most species they were usually included as a family in the Hypocreales in the past. Most species are parasitic on grasses, insects, spiders, and other fungi.

Claviceps — In this genus the perithecial stroma develops as a head on an erect stalk and arises from a dark sclerotium (Plate VII). Examine preserved material of *Claviceps purpurea*, the cause of ergot disease of cereals, on heads of rye; also examine sclerotia bearing perithecial stromata (Plate VII, Fig. 167-168). Look at prepared slides showing sections through young sclerotia bearing the conidial state, *Sphacelia segetum*. Then examine prepared slides of sections through perithecial stromata. Note the perithecia, asci and ascospores (Fig. 167).

Cordyceps — In this genus the stalked perithecial stromata arise from endosclerotia in insects or fungus fruiting bodies. Examine preserved material of *Cordyceps capitata* parasitic on *Elaphomyces*, a hypogeous discomycete, and examine dried material of *Cordyceps clavulata* parasitic on scale insects (Plate VII). Examine prepared slides showing a section through a perithecial stroma.

Balansia — This genus forms black stromata on grasses. Examine stromata on grass leaves. Make thin sections through a stroma, mount, and look for perithecia, asci, and ascospores.

REFERENCES

Diehl, W.W. 1950. *Balansia* and the Balansiae in America. *U.S. Dept. Agr. Monogr. No. 4*: 1-82.

Fuller, John G. 1968. *The Day of St. Anthony's Fire*. New Amer. Lib., Inc., New York, 278 pp.

Jenkins, W.A. 1934. The development of *Cordyceps agariciformia*. *Mycologia* 26: 220-243.

Mains, E.B. 1939. *Cordyceps* from the mountains of North Carolina and Tennessee. *J. Elisha Mitchell Sci. Soc.* 55: 117-129.

Mains, E.B. 1957. Species of *Cordyceps* parasitic on *Elaphomyces*. *Bull. Torrey Bot. Club* 84: 243-251.

Mains, E.B. 1958. North American entomogenous species of *Cordyceps*. *Mycologia* 50: 169-222.

Rykard, D.M., E.S. Luttrell, and C.W. Bacon. 1984. Conidiogenesis and conidiomata in the Clavicipitoideae. *Mycologia* 76: 1095-1103.

Seaver, F.J. 1911. The Hypocreales of North America — IV. *Mycologia* 3: 207-230.

Shanor, L. 1936. The production of mature perithecia of *Cordyceps militaris* (Linn.) Link in laboratory culture. *J. Elisha Mitchell Sci. Soc.* 52: 99-104.

PLATE VII. Upper left, Sclerotia of *Claviceps purpurea* on wheat head. × 2. Upper center, Germinating sclerotium of *Claviceps purpurea,* giving rise to perithecial stromata. × 2. Upper right, Germinating sclerotium of *Claviceps gigantea* with perithecial stromata. × 0.5. Lower left, Perithecial stromata of *Cordyceps capitata* arising from ascocarp of *Elaphomyces* sp. × 2. Lower center, Perithecial stromata of *Cordyceps melolonthae* var. *rickii* arising from insect larva. × 2. Lower right, Sclerotia of *Claviceps gigantea* on ear of corn. × 1.

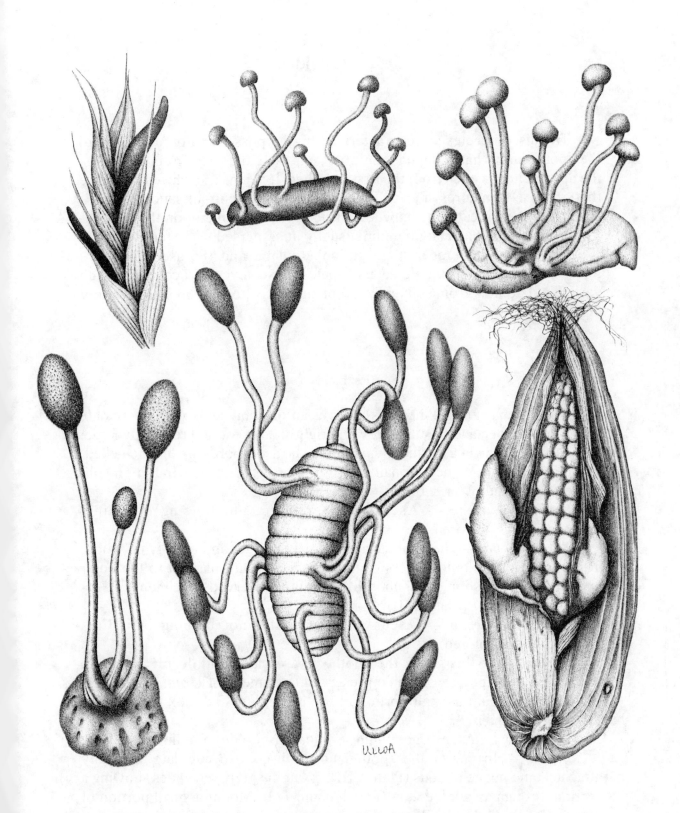

Discomycetidae

The discomycetes are characterized by the possession of an ascocarp, the apothecium, that is open at maturity, exposing the asci. In some species the apothecium is open from the first; in others it begins as a closed structure that opens as it matures. Typically the apothecium is cup- or saucer-shaped, but it may vary greatly in shape and structure, depending upon the species. The asci are borne in an hymenium, usually interspersed with sterile paraphyses. Asci are either operculate or inoperculate, and the ascospores are forcibly discharged in the epigeous species. Discomycetes occur in a wide variety of habitats, from soil and dead plant debris, to parasites on a variety of plants.

Pezizales

The order Pezizales includes all epigean discomycetes with operculate asci. The apothecia are usually fleshy, brightly colored, and relatively large. The asci are borne in a distinct hymenium. The ascospores are all one-celled, hyaline or brown, or occasionally purple. The Pezizales are found on soil, wood, dung, and plant debris.

Sarcoscyphaceae — Ascus suboperculate, not blueing in iodine, apothecia bright-colored.

Sarcoscypha — The apothecia are small to large, and usually on wood. The hyphae of the outer layers of the apothecium are parallel to the outer surface. The ascospores are smooth. Examine preserved specimens of *Sarcoscypha coccinea* (Plate VIII).

Sarcosomataceae — Ascus suboperculate, not blueing in iodine; apothecia dark-colored.

Urnula — In this genus the apothecia are black and deeply cupulate, with ovoid ascospores. Examine preserved specimens of *Urnula craterium* (Plate VIII). Mount a small portion of the hymenium and look for asci and ascospores (Fig. 169).

Pezizaceae — Ascus operculum terminal, asci blue in iodine.

Peziza — In *Peziza* the apothecium is discoid to cupulate, centrally attached, and not gelatinous (Plate VIII). Examine prepared slides showing a section through an apothecium (Fig. 170 and 172). Mount a small portion of hymenium in Melzer's Reagent and note the blueing of the ascus tip.

Examine cultures of *Chromelosporium ollare* (= *Ostracoderma epigaeum*), the conidial state of *Peziza ostracoderma,* under the dissecting microscope. Mount some conidiophores in water and examine under the microscope for ampullae and conidia (Fig. 173).

Pyronemataceae — Ascus operculum terminal; asci not blueing in iodine, ascospores uninucleate; apothecia usually small.

Aleuria — This genus has brightly colored apothecia bearing hyaline hairs (Plate VIII). Mount a small portion of the hymenium of *Aleuria aurantia* in Melzer's Reagent and examine the asci, ascospores, and paraphyses. Note that the asci do not stain blue. The operculum is often difficult to see, but if you examine a number of asci, you should see one. Note the "reticulate" ascospore wall (Fig. 171).

Otidea — The ascocarp is cupulate to ear-shaped, usually slit down one side, lacking hairs. (Plate VIII). Examine preserved material of available species.

Helvellaceae — Ascus operculum terminal, asci not blueing in iodine; ascospores tetranucleate; apothecia large, usually stalked, often saddle-shaped.

Gyromitra — In this genus the pileus is gyrose, brain-like and stipitate. The ascospores are usually biguttulate. Examine preserved specimens of *Gyromitra esculenta* (Plate VIII).

Helvella — In *Helvella* the ascocarp is cupulate to saddle-shaped, with a stipe. The ascospores usually contain a single guttule, and they may be smooth to warted. Examine the saddle-shaped apothecia of *Helvella lacunosa* (Plate VIII) or similar species.

Morchellaceae — Ascus operculum terminal, asci not blueing in iodine; ascospores with 20-60 nuclei; apothecia large, discoid to stalked and sponge-like.

Morchella — The ascocarp is stipitate, with a hollow, sponge-like pileus. Examine preserved specimens of *Morchella esculenta,* the common morel (Plate VIII). Look at prepared slides showing a section through the hymenium (Fig. 175).

REFERENCES

Akechi, K. 1965. Some morphological and ecological studies on *Pyronema domesticum* collected in Japan. *Trans. Mycol. Soc. Japan* 5:37-43.

Bistis, G. 1956. Sexuality in *Ascobolus stercorarius*. I. Morphology of the ascogonium; plasmogamy; evidence for a sexual hormonal mechanism. *Amer. J. Bot.* 43:389-394.

Kimbrough, J.W. 1969. North American species of *Thecotheus* (Pezizeae, Pezizaceae). *Mycologia* 61:99-114.

Kimbrough, J.W. 1970. Current trends in the classification of Discomycetes. *Bot. Rev.* 36:91-161.

Kimbrough, J.W., and R.P. Korf. 1967. A synopsis of the genera and species of the tribe Theleboleae (=Pseudoascoboleae). *Amer. J. Bot.* 54:9-23.

Kimbrough, J.W., E.R. Luck-Allen, and R.F. Cain. 1972. North American species of *Coprotus* (Thelebolaceae: Pezizales). *Canad. J. Bot.* 50:957-971.

Kish, L.P. 1974. Culture and cytological development of *Coprotus lacteus* (Pezizales). *Mycologia* 66:422-435.

Korf, R.P. 1958. Japanese Discomycete notes I-VIII. *Sci. Rept. Yokohama Natl. Univ.* 2(7):7-35.

Korf, R.P. 1972. Synoptic key to the genera of the Pezizales. *Mycologia* 64:937-994.

O'Donnell, K.L., and G.R. Hooper. 1974. Scanning ultrastructural ontogeny of paragymnohymenial apothecia in the operculate discomycete *Peziza quelepidotia*. *Canad. J. Bot.* 52:873-876.

O'Donnell, K.L., W.G. Fields, and G.R. Hooper. 1974. Scanning ultrastructural ontogeny of cleistohymenial apothecia in the operculate discomycete *Ascobolus furfuraceus*. *Canad. J. Bot.* 52:1653-1656.

Paden, J.W. 1972. Imperfect states and the taxonomy of the Pezizales. *Persoonia* 6:405-414.

Paden, J.W. 1973. The conidial state of *Peziza ammophila*. *Canad. J. Bot.* 51:2251-2252.

Paden, J.W., and E.A. Stanlake. 1973. Ascocarp development in *Ascobolus michaudii*. *Canad. J. Bot.* 51:1271-1273.

Rifai, M.A. 1968. The Australasian Pezizales in the herbarium of the Royal Botanic Gardens Kew. *Verhd. Nederl. Akad. Wetensch. Natuurk.* 57(3):1-295.

Seaver, F.J. 1942. *The North American cup-fungi (Operculates).* Suppl. Ed. Pub. by author, New York.

Weber, N.S. 1972. The genus *Helvella* in Michigan. *Mich. Bot.* 11:147-201.

Wolf, F.A. 1958. The conidial stage of *Lamprospora trachycarpa* (Currey) Seaver. *J. Elisha Mitchell Sci. Soc.* 74:163-166.

PLATE VIII. Discomycete apothecia. Top row, l-r. *Urnula craterium.* × 1. *Sarcoscypha coccinea.* × 1. Second row, l-r. *Otidea concinna.* × 1. *Gyromitra esculenta.* × 0.5. *Helvella lacunosa.* × 1. Bottom row, upper left, *Peziza varia.* × 1. Lower left, *Aleuria aurantia.* × 1. Lower right, *Morchella esculenta.* × 1.

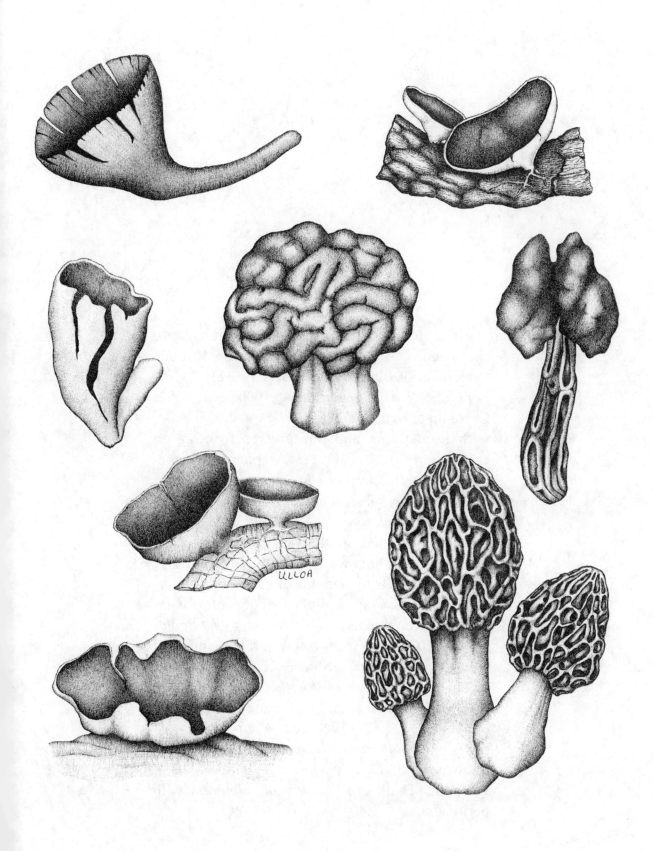

FIG. 169. Asci and ascospores of *Urnula craterium*. × 250. FIG. 170. Section through hymenium of *Peziza* sp. × 50. FIG. 171. Ascospore of *Aleuria aurantia*. × 1000. FIG. 172. Close-up of hymenium of *Peziza* sp. showing asci with ascospores and paraphyses forming an epithecium. × 250. FIG. 173. Ampullae with conidia on conidiophore of *Chromelosporium ollare* (= *Ostracoderma epigaeum*), conidial state of *Peziza ostracoderma*. × 400. FIG. 174. Section through unopened apothecium of *Pseudopeziza medicaginis* in host leaf (HL). × 320. FIG. 175. Section through hymenium of *Morchella* sp. × 250. FIG. 176. Chains of blastospores of *Monilia fructicola*, conidial state of *Monilinia fructicola*. × 400. FIG. 177. Section through hymenium of *Geoglossum* sp. × 100. FIG. 178. Section through apothecium of *Rhytisma acerinum*. × 155. FIG. 179. Asci and ascospores in ascocarp of *Tuber* sp. × 625. FIG. 180. Ascus and ascospores of *Geoglossum difforme*. × 500. FIG. 181. Ascospore of *Geoglossum difforme*. × 1000. FIG. 182. Thallus of *Laboulbenia elongata* with perithecium. × 200.

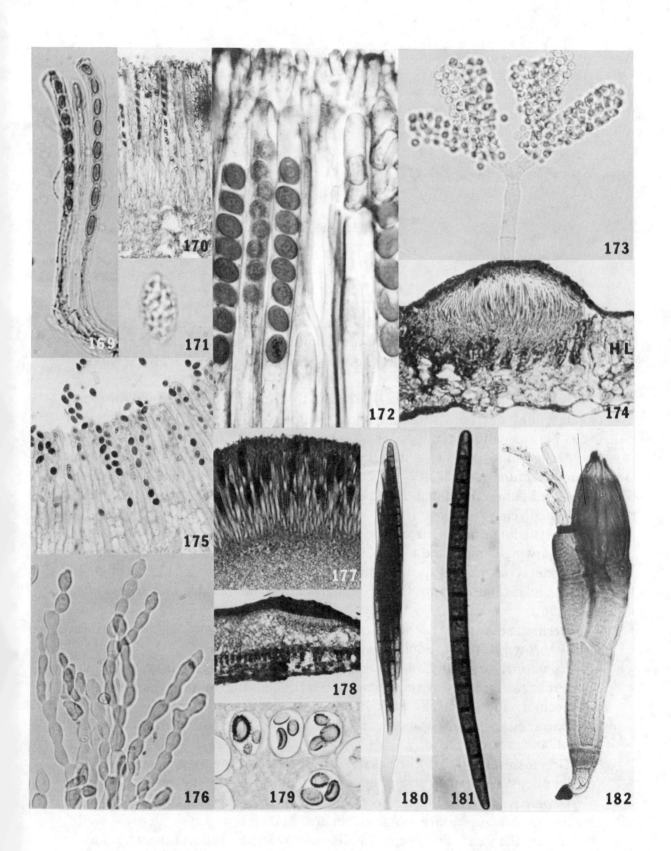

Helotiales

The order Helotiales contains inoperculate discomycetes with forcible spore discharge in which the apothecia are not formed in a stroma. The asci are more or less clavate and have an apical pore through which the spores are discharged. They are usually found as saprobes or plant parasites, rarely on soil or dung.

Sclerotiniaceae — Apothecia cupulate to discoid, arising from a stroma or sclerotium, or from stromatized host tissue, usually brownish and stalked; ascus pore usually blue.

Monilinia — In this genus the apothecia arise from stromatized host tissues or mummified fruits, the conidial state is a *Monilia,* and the ascospores are hyaline. Examine dried peaches (brown rot of peach) mummied by *Monilinia (Sclerotinia) fructicola;* note the clumps of conidial hyphae. Look at preserved mummies bearing apothecia (Plate IX). Examine prepared slides showing conidia and a section through an apothecium. Note the different tissue types in the apothecium. Mount conidia from a culture of this fungus and examine them. The conidial state is named *Monilia fructicola* (Fig. 176).

Dermateaceae — Apothecia cupulate to discoid, erumpent to sessile, usually brown to black, excipulum cells globose to angular.

Pseudopeziza — In *Pseudopeziza* the apothecium arises from a stroma on herbaceous plants, excipulum is nearly lacking at the sides of the apothecium, and the ascus pore blues in iodine. Examine prepared slides showing sections through apothecia of *Pseudopeziza medicaginis* on alfalfa (Fig. 174).

Geoglossaceae — Apothecia clavate, spathulate, or stalked with an irregular pileus, without an abrupt edge to hymenium where it meets the stipe.

Geoglossum — In this genus the apothecia are clavate, the ascospores are brown, and there are no setae in the hymenium. Look at preserved specimens, then examine prepared slides showing a section through an ascocarp (Plate IX). Mount a small portion of the hymenium of *Geoglossum difforme* in water and look for asci and ascospores (Fig. 180-181).

Spathularia — In this genus the apothecia are spathulate, with hymenium on both sides. The ascospores are hyaline. Examine preserved specimens of *Spathularia flavida* (Plate IX).

Leotiaceae — Apothecia cupulate, or clavate or pileate but then with an abrupt edge between hymenium and stipe, fleshy, excipulum usually of long-celled hyphae, not arising from stromatized tissues.

Calycella — In this genus the apothecia are small, sessile to substipitate, and some shade of yellow or orange. The ascospores are hyaline and often one-septate. Examine preserved specimens of *Calycella citrina* (Plate IX); this species was formerly placed in the genus *Helotium*.

Leotia — In *Leotia* the ectal excipulum consists of two layers, the outermost of which is gelatinous. The ascocarp is distinctly capitate, and the ascospores are more than 15 μm long. Look at preserved specimens of *Leotia lubrica*. Mount a small portion of hymenium and look for asci and ascospores (Plate IX).

REFERENCES

Drayton, F.L. 1934. The sexual mechanism of *Sclerotinia gladioli*. *Mycologia* 26: 46-72.

Drayton, F.L. 1934. *The Gladiolus dry rot caused by Sclerotinia gladioli* (Massey) n. comb, *Phytopathology* 24: 397-404.

Hawthorne, B.T. 1973. Production of apothecia of *Sclerotinia minor*. *N.Z.J. Agr. Res.* 16: 559-560.

Honey, E.E. 1928. The monilioid species of *Sclerotinia*. *Mycologia* 20: 127-157.

Honey, E.E. 1936. North American species of *Monilinia*. I. Occurrence, grouping and life-histories. *Amer. J. Bot.* 23: 100-106.

Kohn, L.M. 1979. Delimitation of the economically important plant pathogenic *Sclerotinia* species. *Phytopathology* 69: 881-886.

Korf, R.P., and K.P. Dumont. 1972. *Whetzelinia,* a new generic name for *Sclerotnia sclerotiorum* and *S. tuberosa. Mycologia* 64: 248-251.

Mains, E.B. 1954. North American species of *Geoglossum* and *Trichoglossum*. *Mycologia* 46: 586-631.

Mains, E.B. 1955. North American hyaline-spored species of the Geoglossaceae. *Mycologia* 47: 846-877.

Mains, E.B. 1956. North American species of the Geoglossaceae. Tribe Cudonieae. *Mycologia* 48: 694-710.

Seaver, F.J. 1951. *The North American cup-fungi (Inoperculates), with supplement.* Pub. by author, New York.

Whetzel, H.H. 1945. A synopsis of the genera and species of the Sclerotiniaceae, a family of stromatic inoperculate Discomycetes. *Mycologia* 37: 648-714.

Whetzel, H.H. 1946. The cypericolous and junicolous species of *Sclerotinia*. *Farlowia* 2: 385-437.

Whetzel, H.H., and N.F. Buchwald. 1936. North American Species of *Sclerotinia* and related genera. III. *Ciboria acerina. Mycologia* 28: 514-527.

Ostropales

Apothecia do not develop from a stroma; asci are cylindrical, long, with strongly thickened apex, at least when young. The ascospores are hyaline, filiform, and multiseptate. There is a single family, the **Stictidaceae.**

Vibrissea — Species in this genus form stalked, capitate apothecia on water-logged plant debris. Examine preserved material of *Vibrissea truncorum* (Plate IX).

REFERENCES

Seaver F.J. 1951. The North American cup-fungi (Inoperculates), with supplement. Pub. by author, New York.

Sherwood, M.A. 1977. The Ostropalean fungi. *Mycotaxon* 5: 1-277.

PLATE IX. Discomycete apothecia. Upper left, *Monilinia fructicola.* × 0.5. Upper center, *Calycella citrina.* × 5. Upper right, *Geoglossum difforme.* × 1. Center, *Spathularia flavida.* × 1. Left center, *Leotia lubrica.* × 1. Lower left, *Tuber* sp. × 1. Lower right, *Vibrissea truncorum.* × 5.

Phacidiales

In this order the apothecia develop from a stroma that is usually at least partly immersed in plant tissues. The stroma is usually black externally, variable in shape, and eventually ruptures by a slit at maturity to expose the asci. The asci are thickened apically.

Rhytismataceae — The ascospores are surrounded by a gelatinous sheath.

Rhytisma — In *Rhytisma* the stroma has many hymenial areas and several slits, and the ascospores are hyaline and filiform. Examine dried material of *Rhytisma acerinum* on maple leaves with the dissecting microscope and note the "wrinkled" surface of the stroma. The asci form beneath these wrinkles. Examine a prepared slide showing a section through an ascocarp (Fig. 178).

Hypoderma — The apothecia in *Hypoderma* are elongate, with a single slit, mostly on leaves and stems, and the ascospores are bacillar. Examine infected pine needles under the dissecting microscope and look for the elongate ascocarps characteristic of this genus.

REFERENCES

Darker, G.D. 1932. The Hypodermataceae of Conifers. *Contrib. Arnold Arb.* 1:1-131.

Darker, G.D. 1967. A revision of the genera of the Hypodermataceae. *Canad. J. Bot.* 45:1399-1444.

Duravetz, J.S., and J.F. Morgan-Jones. 1971. Ascocarp development in *Rhytisma acerinum* and *R. punctatum*. *Canad. J. Bot.* 49:1267-1272.

Tehon, L.R. 1935. A Monographic Rearrangement of *Lophodermium*. *Univ. Illinois Bull.* 32:1-150.

Tuberales

In the Tuberales the closed ascocarps are hypogeous and the ascospores are not forcibly discharged. The asci are scattered or arranged in an hymenium, and are cylindrical to spherical. There are eight or fewer ascospores per ascus and the number may vary among asci in the same ascocarp. The ascospores are one-celled and hyaline or brown.

Tuberaceae — Ascocarps hollow to chambered, asci persistent, irregularly arranged or in a distinct hymenium.

Tuber — In *Tuber* there is no hymenium, the ascocarp cavities are filled with hyphae, and the ascospores are sculptured. This is the truffle, prized for eating. Look at preserved specimens and examine a prepared slide showing a section through an ascocarp (Plate IX, Fig. 179).

REFERENCES

Burdsall, H.H., Jr., 1968. A revision of the genus *Hydnocystis* (Tuberales) and of the hypogeous species of *Geopora* (Pezizales). *Mycologia* 60:496-525.

Gilkey, H.M. 1916. A revision of the Tuberales of California. *Univ. Calif. Pub. Bot.* 6:275-356.

Gilkey, H.M. 1939. Tuberales of North America. *Ore. St. Monogr. Stud. Bot.* 1:1-63.

Gilkey, H.M. 1954. Tuberales. *North Amer. Flora, II,* 1:1-36.

Uecker, F.A. 1967. *Stephensia shanori*. I. Cytology of the ascus and other observations. *Mycologia* 59:819-832.

Laboulbeniomycetes

The Laboulbeniomycetes are minute ectoparasites of insects or other arthropods. The thallus, or receptacle, is attached to the host by a basal holdfast from which the haustorium forms. The cells of the receptacle are usually arranged in definite rows. The receptacle bears simple or branched appendages and also bears the perithecium. Thalli may be either monoecious or dioecious. The appendages may be sterile or may form spermatia externally or in simple or compound antheridia. Perithecia bear a simple or branched trichogyne. The unitunicate asci are usually four-spored; the ascospores are two-celled, elongate, more or less fusiform, and surrounded by a gelatinous sheath. There is a single order, the **Laboulbeniales**.

Laboulbeniaceae — Spermatia endogenous, formed in simple antheridia.

Laboulbenia — In *Laboulbenia* the thallus is monoecious and the subbasal cell of the receptacle subtends two cells placed side by side. Examine the demonstration slide of *Laboulbenia elongata* and observe the receptacle, asci, and ascospores (Fig. 182).

REFERENCES

Benjamin, R.K. 1955. New genera of Laboulbeniales. *Aliso* 3: 183-197.

Benjamin, R.K. 1971. Introduction and supplement to Roland Thaxter's Contribution towards a monograph of the Laboulbeniaceae. *Bibliotheca Mycol.* 30: 1-155.

Benjamin, R.K., and L. Shanor. 1950. The development of male and female individuals in the dioecious species *Laboulbenia formicarum* Thaxter. *Amer. J. Bot.* 37: 471-476.

Benjamin, R.K., and L. Shanor. 1951. Morphology of immature stages of *Euzodiomyces lathrobii* Thaxter and the taxonomic position of the genus *Euzodiomyces*. *Amer. J. Bot.* 38: 555-560.

Benjamin, R.K., and L. Shanor. 1952. Sex of host specificity and position specificity of certain species of *Laboulbenia* on *Bembidion picipes*. *Amer. J. Bot.* 39: 125-131.

Shanor, L. 1955. Some observations and comments on the Laboulbeniales. *Mycologia* 47: 1-12.

Tavares, I.I. 1965. Thallus development in *Herpomyces paranensis* (Laboulbeniales). *Mycologia* 57: 704-721.

Tavares, I.I. 1966. Structure and development of *Herpomyces stylopygae* (Laboulbeniales). *Amer. J. Bot.* 53: 311-318.

Tavares, I.I. 1985. Laboulbeniales. *Mycol. Mem.* 9: 1-627.

Thaxter, R. 1895. Contribution toward a monograph of the Laboulbeniaceae. I. *Mem. Amer. Acad. Arts Sci.* 12: 195-429.

Thaxter, R. 1908. Contribution toward a monograph of the Laboulbeniaceae. II. *Mem. Amer. Acad. Arts Sci.* 13: 219-460.

Thaxter, R. 1924. Contribution toward a monograph of the Laboulbeniaceae. III. *Mem. Amer. Acad. Arts Sci.* 14: 309-426.

Thaxter, R. 1926. Contribution toward a monograph of the Laboulbeniaceae. IV. *Mem. Amer. Acad. Arts Sci.* 15: 427-500.

Thaxter, R. 1931. Contribution toward a monograph of the Laboulbeniaceae. V. *Mem. Amer. Acad. Arts Sci.* 16: 1-435.

Loculoascomycetes

The Loculoascomycetes are characterized by having bitunicate asci. The ascus wall consists of two layers, a rigid outer layer, the ectoascus, and a thick, extensible inner layer, the endoascus, which extends through the ectoascus at the time of spore discharge. The ascocarp, or pseudothecium, begins as a stroma (ascostroma) in which the ascogenous system is differentiated. As the asci develop a locule is dissolved around them; ascostromata may be uni- or multi-loculate. In uniloculate pseudothecia the wall of the ascocarp consists of the remains of the stroma following formation of the ascigerous locule. Such ascocarps may closely resemble perithecia of the Euascomycetes, the two being distinguished by the nature of the ascus. In most species the ascospores are forcibly discharged through an ostiole. Ascospores in most species are multicellular and either hyaline or brown.

Myriangiales

This is a relatively small order of species with globose asci individually scattered in the pseudoparenchymatous tissue of the ascostroma. The ascospores are usually phragmosporous or dictyosporous. These fungi are largely tropical or subtropical, and they occur as epiphytes or parasites on superficial fungi or scale insects on living leaves and stems.

Myriangiaceae — Asci distributed irregularly in several layers in the ascostroma.

Myriangium — In *Myriangium* the ascostroma is complex, with fertile and sterile portions. Observe dried material under the dissecting microscope. Examine prepared slides showing sections through an ascostroma, and note the fertile and sterile portions of the ascostroma, the bitunicate asci, and muriform ascospores (Fig. 183).

Elsinoe — The ascostroma is simple, often crust-like, and erumpent in host tissues. The ascospores are phragmosporous and hyaline to brownish. The anamorph is in *Sphaceloma*. Examine leaves of dogwood infected with *Elsinoe corni*.

REFERENCES

Arx, J.A. von. 1963. Die Gattungen der Myriangiales. *Persoonia* 2:421-475.
Miller, J.H. 1938. Studies in the development of two *Myriangium* species and the systematic position of the order Myriangiales. *Mycologia* 30:158-181.
Miller, J.H. 1940. The genus *Myriangium* in North America. *Mycologia* 32: 587-600.

FIG. 183. Section through ascostroma of *Myriangium* sp. showing asci scattered in locules in fertile region of ascostroma. Light region of stroma on right is sterile. × 100. FIG. 184-191. *Sporormiella australis*. FIG. 184. Beginning of ascostroma formation. × 260. FIG. 185. Later stage in ascostroma formation. × 260. FIG. 186. Older ascostroma. × 260. FIG. 187. Mature pseudothecium. × 260. FIG. 188. Ascus with immature ascospores. × 1000. FIG. 189. Mature ascus with ascospores. × 1000. FIG. 190. Expanded bitunicate ascus. Arrow indicates apex of ectoascus. × 500. FIG. 191. Mature ascospore. × 1000. FIG. 192. Section through pseudothecium of *Leptosphaeria* sp. in host leaf (HL). FIG. 193. Section through pseudothecium of *Venturia inaequalis* with asci and pseudoparaphyses. × 500.

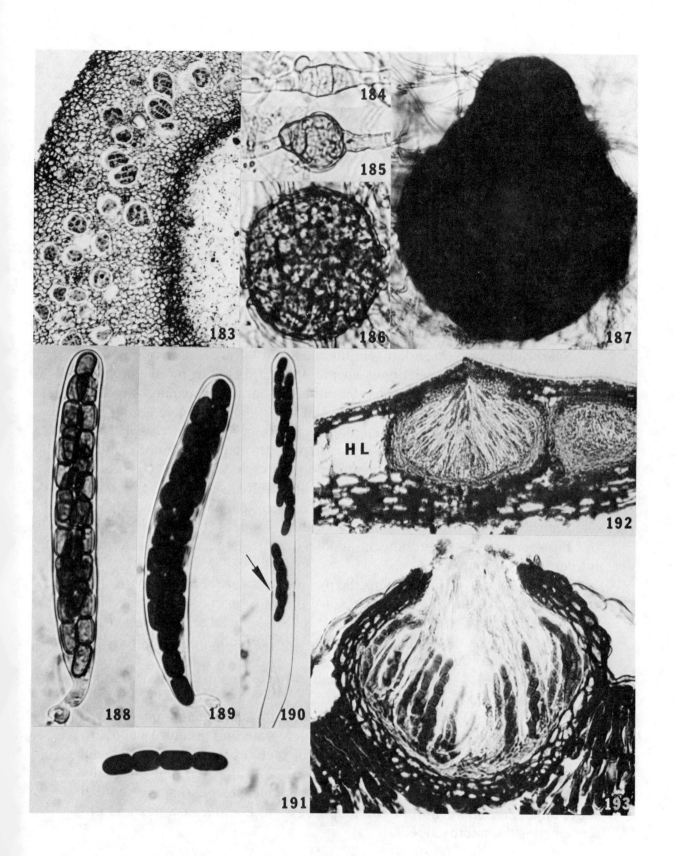

Pleosporales

The Pleosporales are ascostromatic fungi having perithecioid pseudothecia with pseudoparaphyses in the centrum. The ascospores are frequently phragmosporous or dictyosporous. These fungi commonly occur on plant materials, such as dead leaves, herbaceous stems, branches, and wood, as well as on green leaves and stems.

Sporormiaceae — Pseudothecia middle-sized to large, ascospores one- to many-septate, with a gelatinous sheath and germ slits or pores.

Sporormiella — In this genus the phragmosporous ascospores are surrounded by separate gelatinous sheaths and the ascospores have germ slits. Mount mycelium from near the edge of a culture of *Sporormiella australis* and look for different stages in ascostroma formation. Mount mature pseudothecia in water and press gently, do not crush, on the cover glass so as to squeeze out the asci. Look for asci with mature ascospores, which are dark-brown and phragmosporous. Observe ascus elongation. You should also see the slender pseudoparaphyses in these mounts (Fig. 184-191).

Pleosporaceae — Pseudothecia middle-sized to large, ascospores without germ slits or pores.

Leptosphaeria — In this genus the pseudothecia are separate, with very thick-walled cells, and the ascospores are three- to many-septate and concolorous. Examine prepared slides of sections through a leaf bearing pseudothecia. Note the pseudoparaphyses. Some young asci and a few ascospores may also be present (Fig. 192).

Venturiaceae — Pseudothecia small, ascospores with thin, smooth walls, one-septate.

Venturia — In *Venturia* the pseudothecia are separate and immersed in the substrate; the ascospores are unequally septate and greenish-yellow to pale olive brown. Examine prepared slides of sections through pseudothecia of *Venturia inaequalis,* cause of apple scab. Note the asci and unequally two-celled ascospores (Fig. 193). Examine apple leaves bearing the conidial state, *Spilocaea pomi*. Mount conidiophores and conidia in water and observe under the microscope.

Apiosporina — In this genus the pseudothecia are dark and densely crowded on an extensive superificial mycelium on hypertrophied host tissue; the ascospores are septate near the lower end. Examine specimens of *Apiosporina morbosa* (= *Dibotryon morbosum*), cause of black knot disease, and note the effect on the host. The ascigerous lobes can be seen under the dissecting microscope. Examine prepared slides showing sections through the multiloculate ascostroma and note the relation of the stroma to the wood. Examine a locule containing asci and ascospores (Fig. 194). If available, examine a culture of the conidial state, *Cladosporium morbosum,* and mount some conidia in water and observe under the microscope.

REFERENCES

Ahmed, S.I. 1972. Revision of the genera *Sporormia* and *Sporormiella*. *Canad. J. Bot.* 50: 419-477.

Barr, M.E. 1968. The Venturiaceae in North America. *Canad. J. Bot.* 46: 799-864.

Blanchard, R.O. 1972. Ultrastructure of ascocarp development in *Sporormia australis. Amer. J. Bot.* 59: 537-548.

Blancard, R.O. 1972. Origin and development of ascogenous hyphae and pseudoparahyses in *Sporormia australis. Canad. J. Bot.* 50: 1725-1729.

Sivanesan, A. 1977. *The Taxonomy and Pathology of Venturia Species.* J. Cramer, Vaduz.

Wainwright, S.H., and F.H. Lewis, 1970. Developmental morphology of the black knot pathogen on plum. *Phytopathology* 60: 1238-1244.

Wehmeyer, L.E. 1961. *A world monograph of the genus Pleospora and its segregates.* Univ. Michigan Press, Ann Arbor.

Dothideales

In the Dothideales the pseudothecia are generally small and may be multiloculate or perithecioid. The centrum is composed of pseudoparenchyma tissue, and small locules form as the asci develop. The asci are also small and obclavate to short cylindrical. These fungi are very common on dead leaves and stems.

Pseudosphaeriaceae — Pseudothecium perithecioid, immersed, containing a few, relatively large, erect asci.

Leptosphaerulina — This genus is characterized by having perithecioid pseudothecia in dead leaves or stems, and hyaline, dictyosporous ascospores. Mount several pseudothecia of *Leptosphaerulina crassiasca* in water and flatten gently to squeeze out the asci. Observe the globose, bitunicate asci and muriform ascospores (Fig. 195-198).

Dothideaceae — Ascostroma perithecioid or pulvinate, frequently erumpent in the substrate, with asci borne in a fascicle.

Mycosphaerella — Pseudothecia perithecioid, immersed in host tissue, ascospores hyaline and two-celled. Examine prepared slides of *Mycosphaerella* spp. showing sections through pseudothecia.

REFERENCES

Barr, M.E. 1958. Life history studies of *Mycosphaerella tassiana* and *M. typhae*. *Mycologia* 50: 501-513.

Barr, M.E. 1972. Preliminary studies on the Dothideales in temperate North America. *Contrib. Univ. Mich. Herb.* 9: 523-638.

Graham, J.H., and E.S. Luttrell. 1961. Species of *Leptosphaerulina* on forage plants. *Phytopathology* 51: 680-693.

Jackson, C.R., and D.K. Bell. 1968. *Leptosphaerulina crassiasca* (Sechet) comb. nov., the cause of leaf scorch and pepper spot of peanut. *Oleagineux* 23: 387-388.

Muhunya, D.M., and C.W. Boothroyd. 1973. *Mycosphaerella zeae-maydis* sp. n., the sexual stage of *Phyllosticta maydis*. *Phytopathology* 63: 529-532.

Hysteriales

In the Hysteriales the pseudothecia are boat-shaped to linear and carbonaceous. At maturity they open by a slit and become apothecial. The centrum is pseudoparaphysate, with long cylindrical asci. These fungi are common on dead wood.

Hysteriaceae — Pseudothecia elongate, sometimes branched, carbonaceous, saprobic on wood.

Hysterographium — In this genus the lips of the opening are depressed and the ascospores are brown and dictyosporous. Examine twigs with a dissecting microscope and look for elongate ascocarps with a slit. Make cross sections of an ascocarp with a razor blade and mount in water. Look for asci and ascospores (Fig. 199).

REFERENCES

Luttrell, E.S. 1953. Development of the ascocarp in *Glonium stellatum*. *Amer. J. Bot.* 40:626-633.

Zogg, H. 1962. Die Hysteriaceae s. str. und Lophiaceae. *Beitrage zur Kryptogamenflora der Schweiz* 11(3):1-190.

FIG. 194. Section through pseudothecium of *Apiosporina morbosa* showing asci. x 500. FIG. 195-198. *Leptosphaerulina crassiasca.* FIG. 195. Mature pseudothecium. x 100. FIG. 196. Young bitunicate ascus. x 500. FIG. 197. Expanded bitunicate ascus. x 500. FIG. 198. Mature ascospore. x 1000. FIG. 199. Ascospore of *Hysterographium* sp. x 800. FIG. 200. Section through pycnia and aecia of *Puccinia graminis* f. sp. *tritici* on barberry leaf (HL). x 156. FIG. 201. Teliospores of *Uromyces phaseoli*. x 400. FIG. 202. Teliospores of *Gymnosporangium juniperi-virginianae*. x 400. FIG. 203-205. *Puccinia graminis* f. sp. *tritici*. FIG. 203. Section through telium. x 320. FIG. 204. Section through telium x 156. FIG. 205. Mature teliospore. x 400. FIG. 206. Aeciospores of *Cronartium fusiforme.* x 400. FIG. 207. Aeciospores of *Coleosporium solidaginis.* x 400. FIG. 208. Teliospores of *Ustilago maydis.* x 209. Teliospores of *Ustilago nuda.* x 1000.

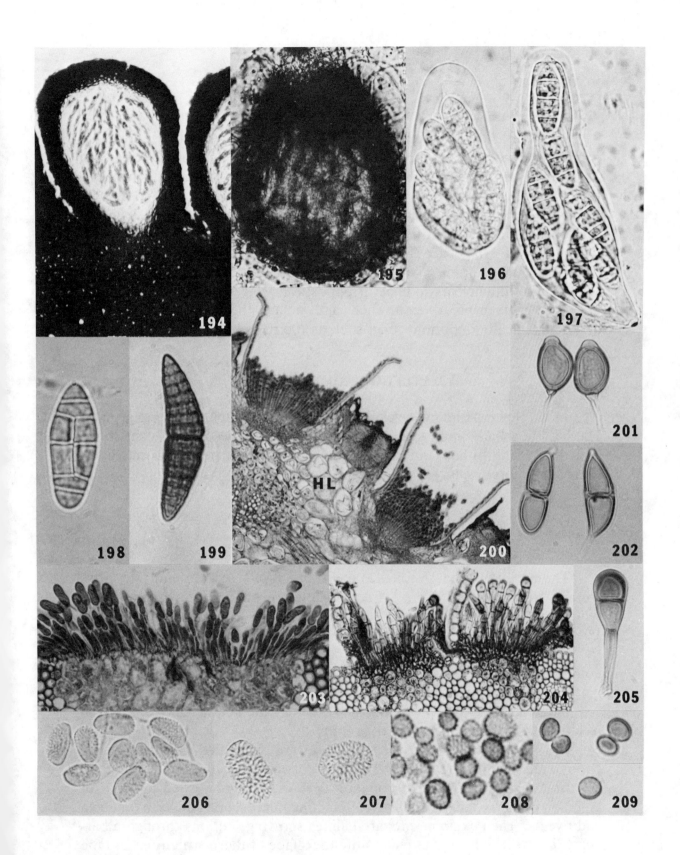

Basidiomycotina

In the Basidiomycotina the perfect state spores (basidiospores) are formed externally on a specialized cell, the basidium. In most species each basidium bears four basidiospores. Basidiospores are forcibly discharged (ballistospores) in most basidiomycetes, and the basidia are commonly borne on some type of basidiocarp. A well developed, septate mycelium is produced by most basidiomycetes. The septum may be simple or dolipore, depending upon the order. The cells of the vegetative mycelium and fructifications are most often dikaryotic, this condition resulting from the anastomosing of uninucleate hyphae of opposite mating types. The basidiomycetes are of widespread occurrence, and include the common mushrooms and economically important fungi such as the rusts and smuts.

Heterobasidiomycetes

In the Heterobasidiomycetes the basidia vary in shape but all are septate. The basidia usually arise from a thick-walled cell. The class contains both species with and without fruiting bodies. The vegetative mycelium may have simple or dolipore septa, depending upon the group.

Uredinales

The basidia in the Uredinales are transversely septate and they arise from a thick-walled teliospore. Each basidium produces four basidiospores which are forcibly discharged. No fruiting body is formed. Species in the Uredinales have the most complex life cycles among the fungi, with up to five separate spore stages. They are commonly called rust fungi, from the color of certain spore stages on the host. All of the rust fungi are obligate parasites of vascular plants. The mycelium is simple septate.

Pucciniaceae — The teliospores are stalked and bear a septate, external basidium. They are never united laterally.

Puccinia — In *Puccinia* the teliospores are two-celled, pedicellate, and colored, and the pycnia are globose and subepidermal. *Puccinia graminis* f. sp. *tritici* is one of the most common species. Examine dried material of common barberry leaves under the dissecting microscope and look for aecia and pycnia. Then examine prepared slides showing sections through a barberry leaf with aecia and pycnia. Note the shape of these structures and the spores they bear. Examine wheat stems under the dissecting microscope and look for uredinia and telia. Mount some urediniospores and teliospores in water and observe under the microscope. Examine prepared slides showing sections through wheat stems bearing uredinia and telia (Fig. 200, and 203-205).

Uromyces — *Uromyces* differs from *Puccinia* in that it has one-celled teliospores. Mount some teliospores of *Uromyces phaseoli* from cowpea leaves in water and observe under the compound microscope (Fig. 201).

Gymnosporangium — In this genus the telia are gelatinous and they occur on Cupressaceae. *Gymnosporangium juniperi-virginianae*, cause of cedar apple rust, is common on red cedar. Examine leaves of hawthorn for pycnia and aecia, on either dried or preserved material. Examine freeze-dried telial galls on red cedar; mount a portion of a telial column in water and examine the teliospores under the microscope (Fig. 202).

Gymnosporangium nidus-avis forms telial masses on red cedar stems. Examine preserved material of this stage.

Melampsoraceae — The one-celled teliospores are sessile and produce an external basidium, but they are united laterally into a crust.

Cronartium — This genus has tall, hairlike columns of teliospores which are one-celled and are arranged in vertical chains without intercalary cells. *Cronartium fusiforme* is the cause of fusiform rust of pine. Look at preserved or dried aecial galls on pine. Mount some aeciospores in water and examine under the microscope (Fig. 206). Examine oak leaves bearing telial columns under the dissecting microscope. Mount a portion of a telial column in water and examine under the microscope. Some workers regard this fungus as a forma specialis of *C. quercuum*.

Coleosporiaceae — The teliospores are sessile and are united laterally. No external basidium is formed; the teliospores become septate and function as basidia.

Coleosporium — In *Coleosporium* the one-celled teliospores are arranged in flat, gelatinous telia, and at maturity the teliospores become septate, to form internal basidia. *Coleosporium solidaginis* is a common species. Look at preserved material showing aecia on pine needles and telia on goldenrod. Mount aecia from dried pine needles and look for aeciospores (Fig. 207).

REFERENCES

Arthur, J.C. 1934. *Manual of the rusts of the United States and Canada.* Purdue Res. Foundation. Reprinted with a supplement, 1962, Hafner Pub. Co., New York.

Cummins, G.B. 1971. *The Rust Fungi of Cereals, Grasses, and Bamboos.* Springer-Verlag, Berlin.

Cummins, G.B., and Y. Hiratsuka. 1983. *Illustrated Genera of Rust Fungi.* (Rev. ed.). Amer. Phytopathol. Soc., St. Paul.

Kern, F.D. 1973. *A Revised Taxonomic Account of Gymnosporangium.* Pennsylvania State Univ. Press, University Park.

Littlefield, L.J. 1981. *Biology of the Plant Rusts. An Introduction.* Iowa State Univ. Press, Ames.

Wilson, M., and D.M. Henderson. 1966. *British Rust Fungi.* Cambridge University Press. Cambridge.

Ziller, W.G. 1974. *The Tree Rusts of Western Canada.* Canad. For. Serv. Publ. No. 1329, Victoria.

Ustilaginales

The basidia are transversely septate or non-septate, and arise from intercalary teliospores. The basidiospores are produced in infinite numbers and are not forcibly discharged. Both a haploid and a diploid mycelium is produced; the former will grow in culture as a yeast-like colony, but the latter is obligately parasitic on angiosperms. The teliospores form in sori in various host tissues, but they are especially common in the infloresence. No fruiting body is formed. The teliospores may form singly or in balls; teliospore balls may consist entirely of fertile cells or they may also contain sterile cells. The large numbers of dark-colored teliospores formed on the host gives rise to the common name of smut fungi for these organisms.

Ustilaginaceae — The basidia are septate and basidiospores are produced laterally from each cell.

Ustilago — In this genus the mature sorus is dusty, the teliospores are small, and the sorus lacks a peridium. Examine prepared slides showing sections through oat flowers infected with *Ustilago avenae*. Examine corn infected with *Ustilago maydis* (Plate X). Mount some teliospores in water and examine under the microscope (Fig. 208). Examine barley heads infected with *Ustilago nuda* (Fig. 209).

Tilletiaceae — The basidium is non-septate and the basidiospores are terminal on the basidium.

Tilletia — In *Tilletia* the sori are dusty at maturity, but the spores are small. Examine wheat heads infected with *Tilletia caries*. Mount some teliospores in water and examine under the microscope.

REFERENCES

Fischer, G.W. 1953. *Manual of North American smut fungi.* Ronald Press, New York.

Fischer, G.W., and C.S. Holton. 1957. *Biology and control of the smut fungi.* Ronald Press, New York.

PLATE X. Upper left, Teliospore galls of *Ustilago maydis* on ear of corn. × 1. Upper right, top. Basidiocarp of *Tremella mesenterica*. × 1. Bottom of branch, Basidiocarps of *Exidia glandulosa*. × 1. Lower left, Basidiocarp of *Dacrymyces palmatus*. × 1. Center, Basidiocarps of *Calocera viscosa*. × 2. Lower center, Colony of *Septobasidium burtii*. × 1. Lower right, Basidiocarps of *Auricularia auricula*. × 1.

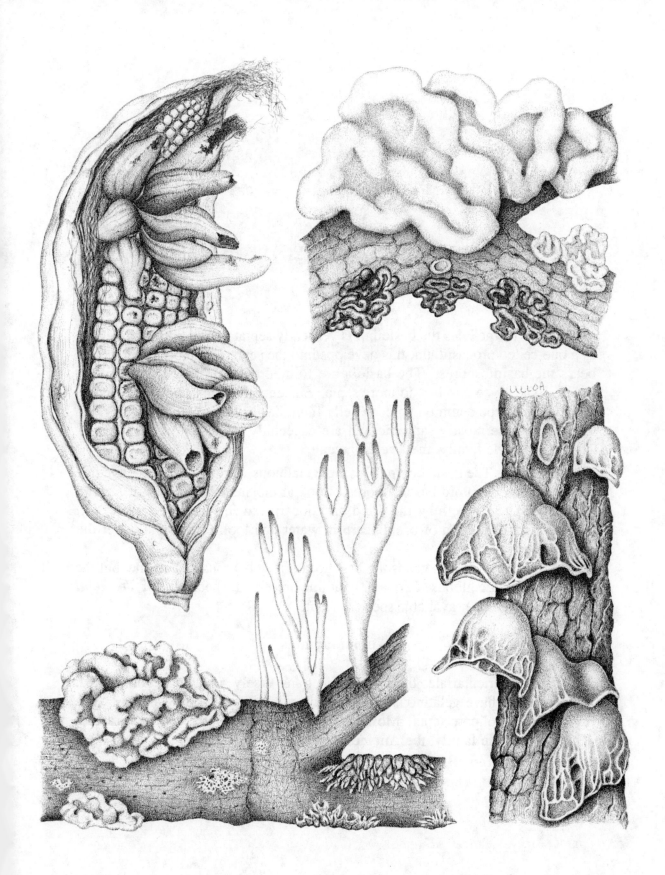

Tremellales

In the Tremellales the basidium is vertically septate. The basidium begins as a one-celled probasidium; this develops into the septate metabasidium which bears the basidiospores. The basidia are formed on well-developed fruiting bodies. The mycelium has dolipore septa. Since many species have gelatinous fructifications the common name of jelly fungi has been applied to them. The Tremellales are mostly saprobic and are especially common on dead wood. There is a single family, the **Tremellaceae.**

Exidia — The basidiocarp is sessile, gelatinous, variable in shape, and has cylindrical to allantoid basidiospores. Look at moistened branches of *Exidia glandulosa* and note how the dried basidiocarps swell when wet (Plate X)). Mount a small portion of basidiocarp in water and look for basidia under the microscope.

Tremella — In this genus the basidiocarp is gelatinous and sessile, but the basidiospores are globose to ovate. Examine preserved specimens of *Tremella mesenterica* or other available species. (Plate X).

Auriculariales

In the Auriculariales the basidium is transversely septate and four-celled. Some species have gelatinous fruiting bodies, but in others they are dry. The hyphae have dolipore septa. Most species are saprobic, but a few are parasitic. There is a single family, the **Auriculariaceae.**

Auricularia — The basidiocarp is variable in shape, firm-gelatinous, with unilateral hymenium. Examine a common species, such as *Auricularia auricula,* and note the shape and consistency of the basidiocarp (Plate X). Mount a small portion of a basidiocarp in water and look for basidia under the microscope.

Septobasidiales

In the Septobasidiales the basidium is transversely septate. All are parasitic or symbiotic on scale insects. The fruiting body consists of a dry, often crust-like mycelial mat on the surface of the substrate. This mat covers the scale insects, which are parasitized. There are no clamp connections and the septa are simple. There is a single family, the **Septobasidiaceae.**

Septobasidium — This is the only genus. Examine mycelial mats of available species under the dissecting microscope (Plate X).

REFERENCES

Couch, J.N. 1938. *The genus Septobasidium.* Univ. North Carolina Press, Chapel Hill. 480 pp.

Lowy, B. 1955. Illusrations and keys to the tremellaceous fungi of Louisiana. *Lloydia* 18: 149-181.

Lowy, B. 1971. Tremellales. *Flora Neo-tropica Monogr.* 6: 1-160. Hafner Publ. Co., Inc., New York.

Martin, G.W. 1952. Revision of the North Central Tremellales. *State Univ. Iowa Stud. Nat. Hist.* 19(3): 1-22.

Olive, L.S. 1947. Notes on the Tremellales of Georgia. *Mycologia* 39: 90-108.

Holobasidiomycetes

In the Holobasidiomycetes the basidia are uniformly single clavate cells that bear basidiospores at the apex. In most groups the basidia are clavate but in one order they are bifurcate. Usually four basidiospores are formed; they are forcibly discharged. A definite fruiting body is formed in all but one small group. The vegetative mycelium has dolipore septa and bears clamp connections in many species. Many of these fungi are found on soil, dung, and decaying wood.

Hymenomycetidae

In the hymenomycetes the basidia are borne in an hymenium that is exposed at maturity. The hymenium may be borne free on the host surface, in pores, on lamellae (gills) or teeth, or on the outer surface of the fruiting body. The fruiting body varies from fleshy and short-lived to woody and persistent.

Dacrymycetales

The Dacrymycetales are characterized by the possession of a one-celled, bifurcate basidium that bears two basidiospores. The fruiting bodies are small and gelatinous or waxy and they occur mostly on decaying wood. The hyphae have a dolipore septum. There is a single family, the **Dacrymycetaceae.**

Calocera — In *Calocera* the basidiocarp is cylindrical to clavate, simple or branched, and the hymenium is amphigenous. Examine preserved specimens of *Calocera viscosa* (Plate X) or other common species.

Dacrymyces — The basidiocarp is gelatinous and is composed only of thin walled hyphae. Examine preserved material of a common species, such as *Dacrymyces deliquescens* or *D. palmatus* (Plate X). If available, mount a small portion of a basidiocarp in water and look for basidia.

REFERENCES

Kennedy, L.L. 1958. The genera of the Dacrymycetaceae. *Mycologia* 50: 875-895.
Kennedy, L.L. 1958. The genus *Dacrymyces*. *Mycologia* 50: 896-915.
Kennedy, L.L. 1972. Basidiocarp development in *Calocera cornea*. *Canad. J. Bot.* 50: 413-417.
Reid, D.A. 1974. A monograph of the British Dacrymycetales. *Trans. Brit. Mycol. Soc.* 62: 433-494.

Exobasidiales

The Exobasidiales are characterized by having the basidia borne free on the surface of the host; no basidiocarp is formed. All members of the order are parasitic on the stems, leaves, or flower buds of phanerogams, where they often cause hypertrophy of the infected parts. There is a single family, the **Exobasidiaceae**.

Exobasidium — In this genus the basidia form a continuous hymenium on the infected host tissue. Examine preserved material of *Exobasidium camelliae* on *Camellia* and note the effect on the host. Examine prepared slides of sections through infected camellia leaves and look for basidia (Fig. 210). Examine preserved material of *Exobasidium azaleae* on azalea.

REFERENCES

Graafland, W. 1953. Four species of *Exobasidium* in pure culture. *Acta Bot. Neerl.* 1: 516-522.

Graafland, W. 1960. The parasitism of *Exobasidium japonicum* Shir. on Azalea. *Acta Bot. Neerl.* 9: 347-379.

McNabb, R.F. R. 1962. The genus *Exobasidium* in New Zealand. *Trans. Roy. Soc. N.A., Botany* 1: 259-268.

Mims, C.W. 1982. Ultrastructure of the haustorial apparatus of *Exobasidium camilliae*. *Mycologia* 74: 188-200.

Mims, C.W., and N.L. Nickerson. 1986. Ultrastructure of the host-pathogen relationship in red leaf disease of lowbush blueberry caused by the fungus *Exobasidium vaccinii*. *Canad. J. Bot.* 64: 1338-1343.

Mims, C.W., and E.A. Richardson, and R.W. Roberson. 1987. Ultrastructure of basidium and basidiospore development in three species of the fungus *Exobasidium*. *Canad. J. Bot.* 65: 1236-1244.

Savile, D.B.O. 1959. Notes on *Exobasidium*. *Canad. J. Bot.* 37: 641-656.

Agaricales

The Agaricales have the basidia borne on basidiocarps that are fleshy and decay away after maturity. The basidiocarp is usually capitate, with a stipe, and the basidia are borne on gills, in pores, or rarely on a smooth surface. The basidia are two- to eight-spored, and the basidiospores are forcibly discharged at maturity. The basidiospores are one-celled.

Boletaceae — Hymenium lining pores on the underside of a fleshy pileus.

Boletus — In *Boletus* the hymenophore is distinctly poroid, the stipe is not hollow, and the pileus is not viscid. Examine *Boletus edulis* (Plate XIII) or other available specimens of *Boletus* spp. and note the characteristic features.

Strobilomyces — In this genus the pileus is covered with coarse blackish scales and the basidiospores are blackish-brown, ornamented, and globose to subglobose. Examine preserved specimens of *Strobilomyces floccopus* (Plate XIII).

Russulaceae — Characterized by the presence of sphaerocysts in the pileus (heteromerous), and the hymenium on gills; basidiospores with amyloid ornamentation.

Lactarius — In this genus latex is present in the basidiocarp. Examine preserved specimens of different species (Plate XII).

Russula — Latex is absent from the basidiocarp. Make a water mount of tissue from the pileus of *Russula emetica* and look for sphaerocysts (Fig. 211-212). Mount a portion of a gill and look for basidia. Add a few drops of Melzer's Solution and note the amyloid basidiospore ornamentation.

Agaricaceae — Pileus trama homiomerous; lamellae free, spore deposit blackish to chocolate brown, annulus typically present, spores lacking an apical germ pore.

Agaricus — Spores not olive, and ovoid to ellipsoid and smooth.

Agaricus brunnescens — This is the mushroom grown for commercial use; it is characterized by having two-spored basidia. Examine specimens of various ages and note the angiocarpic development with partial veil. Mount a small portion of a gill from a young sporophore in water and look for basidia under the microscope. This species has also been called *A. bisporus* and *A. campestris* var. *bisporus*.

Agaricus campestris — Examine specimens of this species, the common meadow mushroom. It has four-spored basidia (Plate XIII, Fig. 213).

Amanitaceae — Pileus trama homiomerous, lamellae free, with a divergent trama. Spore deposit white, with smooth spores. An outer and an inner veil are typically present.

Amanita — In this genus a volva is typically present.

Amanita verna, A. virosa. These species are pure white and have a sac-like volva, white annulus, and smooth cap. Both are very poisonous, usually deadly (the destroying angel) (Plate XI).

Amanita muscaria (the fly agaric) has a yellow to reddish cap with veil patches. The volva consists of several concentric rings above the bulb. It is poisonous. Examine preserved specimens (Plate XI).

Amanita umbonata has a large reddish-orange, smooth cap, a large annulus, and a deep, sac-like volva. It is nonpoisonous. Examine preserved specimens (Plate XI). This species was formerly called *Amanita caesarea*.

Cantharellaceae — The trama is homiomerous and the lamellae are decurrent, with blunt edges, sometimes occurring only as broad ridges. The basidia are long and narrow.

Cantharellus — In this genus the tramal hyphae have clamps and the flesh of the sporophore is thick. Examine specimens of *Cantharellus cibarius*, the golden chanterelle; this species has yellow-orange, funnel-shaped fruiting bodies (Plate XIII). It is highly prized for eating.

Craterellus — In *Craterellus* the tramal hyphae lack clamps and the flesh of the sporophore is thin. Examine specimens of *Craterellus cantharellus*. This species resembles *Cantharellus cibarius*, but the flesh is thinner and the gill ridges are usually less pronounced. Edible (Plate XIII).

Coprinaceae — The trama is homiomerous and the lamellae are attached to the stipe and thin. The spore deposit is blackish to dark-brown. The sporophores are usually very fragile.

Coprinus — The pileus and lamellae undergo autodigestion, forming an inky mass. Examine preserved specimens of common species, such as *Coprinus micaceus* (Plate XII) or *C. comatus*, then examine prepared slides showing sections of the gills of *Coprinus* sp. and look for basidia (Fig. 214-215).

Lepiotaceae — The trama is homiomerous, the stipe and pileus are readily separable, and the lamellae are free. The spore deposit is greenish to buff or off-white. An annulus is typically present.

Chlorophyllum — In *Chlorophyllum* the spore print is greenish and there is no volva. Examine specimens of *Chlorophyllum molybdites* (Plate XII), a poisonous species with large scaly caps that occurs as fairy rings in grassy areas. This species was formerly in the genus *Lepiota*.

Lepiota — The spore print is white, there is no volva, and most spores lack a germ pore. Examine specimens of *Lepiota procera* (Plate XII), the parasol mushroom, an edible species with tall, large, scaly sporophores. Young specimens are easily confused with those of *Chlorophyllum molybdites*.

Strophariaceae — The trama is homiomerous, with the gills attached to the stipe. The spore print is purple-brown and the spores have an apical pore.

Psilocybe — The spore deposit is vinaceous to purple-brown and the stipe is central. They are common on soil and dung and several species are prized for their hallucinogenic properties. Examine preserved specimens of available species, such as *Psilocybe aztecorum, P. caerulescens* (Plate XIII), *P. cubensis,* and *P. mexicana,* all of which are hallucinogenic.

Tricholomataceae — The trama is homiomerous and the stipe and gills are attached and often decurrent. The spore deposit is white to yellowish, buff, or vinaceous-brown.

Armillariella — The sporophores are fleshy and easily broken and they occur in caespitose clusters on wood, in which black rhizomorphs are often formed. Examine preserved rhizomorphs, then look at the following species and note the characters indicated.

Armillariella (Armillaria) mellea — This is the honey mushroom, characterized by the presence of an annulus (Plate XIII).

Armillariella (Clitocybe) tabescens — In this species there is no annulus. Look at rhizomorphs of this species produced in culture.

Clitocybe — In *Clitocybe* the basidiocarp is centrally stipitate and the lamellae are adnate to distinctly decurrent. Clamp connections are typically present. The spores are smooth or ornamented. Examine preserved specimens of a common species, such as *Clitocybe dealbata* (Plate XIII).

Marasmius — Members of this genus occur on soil or on plant debris. The sporophores are small, with black to yellowish-brown, pliant stipes and a cellular cuticle (Plate XII). The spores are white and non-amyloid. Examine preserved specimens of available species.

Panus — This genus occurs on wood and the stipe is eccentric or lacking. The cap is smooth, as are the gill edges. Examine specimens of available species (Plate XII).

Pleurotus — Similar to *Panus,* but with larger, fleshier sporophores. Examine specimens of *Pleurotus ostreatus,* the oyster mushroom. This species has gymnocarpic development (Plate XII).

REFERENCES

Bigelow, H.E. 1974. *Mushroom Pocket Field Guide.* Macmillan Publ. Co., Inc., New York.

Chang, S.T., and W.A. Hayes. (Eds.). 1978. *The Biology and Cultivation of Edible Mushrooms.* Academic Press, New York.

Christensen, C.M. 1970. *Common Edible Mushrooms.* Univ. Minnesota Press, Minneapolis.

Hesler, L.R. 1960. *Mushrooms of the Great Smokies.* Univ. Tennessee Press, Knoxville.

Jülich, W. 1981. *Higher Taxa of Basidiomycetes.* J. Cramer, Vaduz.

McKenny, M. 1971. The Savory Wild Mushroom. Revised by D.E. Stuntz. Univ. Washington Press, Seattle.

Miller, O.K., Jr. 1972. *Mushrooms of North America*. E.P. Dutton & Co., Inc., New York.
Miller, O.K., Jr., and H.H. Miller. 1980. *Mushrooms in Color*. E.P. Dutton. New York.
Pacioni, G. 1981. *Simon and Schuster's Guide to Mushrooms*. Simon and Schuster, New York.
Phillips, R. 1981. *Mushrooms and Other Fungi of Great Britain and Europe.*, Pan Books, Ltd., London.
Savonius, M. 1973. *All Color Book of Mushrooms and Fungi*. Crescent Books, New York.
Schwalb, M.N., and P.G. Miles. (Eds.). 1978. *Genetics and Morphogenesis in the Basidiomycetes*. Academic Press, New York.
Singer, R. 1975. *The Agaricales in Modern Taxonomy*. 3rd ed. J. Cramer, Weinheim.
Smith, A.H. 1949. *Mushrooms in their natural habitats*. Vols. 1 and 2. Sawyer's Inc., Portland.
Smith, A.H. 1975. *A Field Guide to the Western Mushrooms*. Univ. Michigan Press, Ann Arbor.
Smith, A.H., H.V. Smith, and N.S. Weber. 1980. *How to know the gilled mushrooms*. Wm. C. Brown Co. Publ., Dubuque.
Smith, A.H., H.V. Smith, and N.S. Weber. 1981. *How to know the non-gilled mushrooms*. 2nd ed. Wm. C. Brown Publ., Dubuque.
Smith, A.H., and N.S. Weber. 1980. *The Mushroom Hunter's Field Guide*. 3rd ed. Univ. Michigan Press, Ann Arbor.
Stubbs, A.H. 1971. *Wild Mushrooms of the Central Midwest*. Univ. Press of Kansas, Lawrence.
Tosco, U. 1973. *The World of Mushrooms*. Crown Publ., Inc., New York.
Tosco, U., and A. Fanelli. 1972. *Mushrooms and Toadstools. How to find and identify them*. Crown Publ., Inc., New York.
Weber, N.S., and A.H. Smith. 1985. *A Field Guide to Southern Mushrooms*. Univ. Michigan Press. Ann Arbor.
Wells, K., and E.K. Wells. (eds.). 1982. *Basidium and Basdiocarp*. Springer-Verlag, New York.

PLATE XI. Mature basidiocarps. Upper left, *Amanita umbonata*. Upper right, *Amanita virosa*. Bottom, *Amanita muscaria*. All x 1.

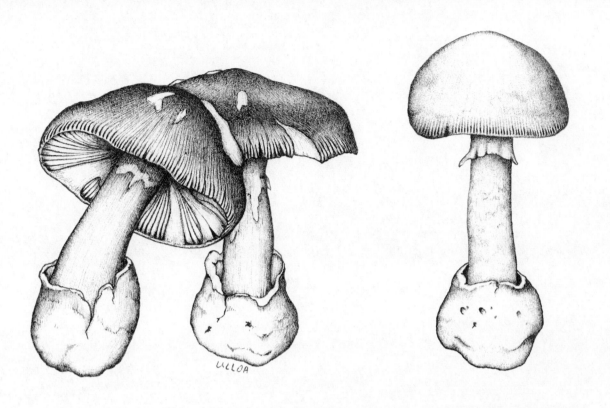

PLATE XII. Mature Basidiocarps. Upper left, *Lepiota procera.* × 1. Upper center, *Russula brevipes.* × 0.5. Upper right, *Lactarius piperatus.* × 1. Middle left, *Chlorophyllum molybdites.* × 1. Middle right, *Pleurotus ostreatus.* × 1. Lower left, *Marasmius siccus.* × 2. Lower right center, *Panus rudis.* × 1. Lower right, *Coprinus micaceus.* × 0.5

PLATE XIII. Mature basidiocarps. Upper left. *Armillariella mellea.* Upper right, *Clitocybe dealbata.* Middle left, *Agaricus campestris.* Center, *Psilocybe caerulescens.* Middle right, *Strobilomyces floccopus.* Lower left, *Cantharellus cibarius.* Lower middle, *Craterellus cornucopioides.* Lower right, *Boletus edulis.* All × 1.

Aphyllophorales

The Aphyllophorales includes those Hymenomycetes which do not have gills on the fruiting body. The basidiocarp is gymnocarpic and varies in form from appressed, to resupinate, pileate, clavate, or coralloid. The hymenial surface may be smooth, plicate, on teeth, or in pores. In many species the fruiting body is persistent and tough or woody. The basidia are arranged in an hymenium and the basidiospores are forcibly discharged. The Aphyllophorales are mostly saprobic on soil, wood, and litter, but some are parasites on trees, on which they form an external basidiocarp.

Clavariaceae — The hymenium is smooth and covers the entire surface of upright fruiting bodies. The fruiting body may be thick and club-shaped, slender and simple, or much branched. These fungi occur on soil and decaying wood.

Clavaria —This genus originally included most of the coralloid fungi, but most species have been transferred to other genera. Examine the prepared slide labelled *Clavaria* and look at sections through a sporophore branch covered with hymenium.

Clavariadelphus — In *Clavariadelphus* there are swellings on the hyphae adjacent to the clamp connections. The sporophore is simple and the spores are smooth and thin-walled. Examine sporophores of a common species, such as *Clavariadephus pistillaris* (Plate XIV).

Clavicorona — In this genus the branches grow upright and the spores are roughened and amyloid. Examine specimens of the common species, *Clavicorona pyxidata* (Plate XIV).

Schizophyllaceae — The basidiocarp is cupulate at first, attached by a narrow base, monomitic, with hyaline spores.

Schizophyllum — In *Schizophyllum* the hymenium is borne on gills which split along their entire length at maturity. Examine specimens of *Schizophyllum commune,* a common species on dead wood, which forms small, sessile, hairy, tough, gray basidiocarps (Plate XIV). If available, examine cultures of this fungus.

Thelephoraceae — The hymenial surface is smooth and is borne on only one side of the sporophore. The sporophore context is pallid to dark. They commonly occur on dead wood.

Corticium — The fruiting body is effuse, with a smooth hymenium that lacks cystidia. Examine specimens of *Corticium coeruleum,* which forms a thin, bright blue basidiocarp on dead wood (Plate XV). Mount a small portion of basidiocarp in water and look for clamp connections (Fig. 224).

Sparassis — In *Sparassis* the basidiocarp is composed of flattened or petaloid branches which bear the hymenium only on one side. Examine specimens of *Sparassis crispa,* the cauliflower fungus (Plate XIV).

Stereum — In this genus the spores are smooth and amyloid, the hyphal system is never trimitic, and there are no clamps. The sporophore may be resupinate or bracket-like. Examine specimens of the following species.

Stereum frustulosum — In this species the basidiocarp is resupinate and often divided up into small cubical or rectangular sections. It occurs on dead wood (Plate XV).

Stereum gausapatum — This has thin, tough sporophores which are attached laterally to the woody substrate.

Hydnaceae — The hymenium is borne on the surface of teeth or tooth-like structures on the basidiocarp. The basidiocarp may be fleshy or woody.

Hericium — The basidiocarps of *Hericium* consist of a system of branches bearing positively geotrophic spines. The context is monomitic and both the context and spores are strongly amyloid. Examine specimens of common species, such as *Hericium erinaceus* and *H. coralloides* (Plate XIV).

Hydnum — In *Hydnum* the basidiocarp is stipitate and the context is fleshy and brittle. The spores are brown and tuberculate. Look at preserved specimens of *Hydnum imbricatum* (Plate XIV). Examine prepared slides showing sections through teeth of *Hydnum* sp.

Dentinum – In this genus the basidiocarp is stipitate and the context is fleshy. The spores are hyaline, smooth, and thin-walled. Examine preserved specimens of *Dentinum repandum* (Plate XIV).

Echinodontium — Species in this genus have perennial, woody basidiocarps that grow on wood. The hymenium is on coarse spines; the spores are hyaline and amyloid. Examine specimens of *Echinodontium tinctorium,* the Indian paint fungus (Plate XV).

Steccherinum — This genus also occurs on wood. The basidiocarp is reflexed-pileate, with a context that does not darken in KOH. The spores are amyloid. Examine specimens of *Steccherinum septentrionale* (Plate XV).

Polyporaceae — The basidiocarp is resupinate to pileate, sessile to stipitate, annual or perennial. The hymenium is poroid from the first and the spores are usually smooth.

Daedalea — In *Daedalea* the hymenophore is labyrinthiform, the hyphal system is trimitic, and the basidiocarp thick and wood-colored to rust brown. Examine specimens of common species, such as *Daedalea quercina* (Plate XV) or *D. ambigua.*

Fomes — The basidiocarp in *Fomes* is perennial, with several layers of tubes. The context is dark from the first and the hyphal system is trimitic, with clamp connections. Examine specimens of common species, such as *Fomes pini* (Plate XV).

Ganoderma — This genus has stipitate basidiocarps, with a tubular hymenophore. The spores are brown, ellipsoid, echinulate, and truncate at one end at maturity. The surface of the sporophore often has a varnished appearance. Examine specimens of common species, such as *Ganoderma curtisii* or *G. tsugae* (Plate XV).

Heterobasidion — The sporophore is perennial and has a context that is white to pale wood color. The hyphal system is dimitic, with skeletal hyphae, and there are no clamp connections. Examine specimens of *Heterobasidion annosum,* cause of root rot of conifers. This fungus has been placed in *Fomes,* as *F. annosus.* Examine wood rotted by *H. annosum;* this is a white rot fungus. Look at mature sporophores. Mount a small portion of a culture in water and look for conidiophores and conidia. The conidial state is *Spiniger meineckellus* (= *Oedocephalum lineatum*) (Fig. 220-223).

Irpex — The basidiocarp is resupinate-reflexed, with a light context that is firm. The hyphal system is dimitic, and the hymenium is borne on toothlike structures. The spores are ovoid to ellipsoid, thin-walled, and inamyloid. Examine specimens of a common species, such as *Irpex cinnamomeus* (Plate XIV).

Lenzites — In *Lenzites* the basidiocarp is annual, with a light context which is firm. The hymenophore is lamellate and the hyphal system is trimitic. The spores are smooth, cylindric, and hyaline. Examine specimens of *Lenzites saepiaria* (Plate XV).

Merulius — In this genus the basidiocarp is effused-reflexed to pileate and the hymenophore is smooth, reticulate, poroid, or irregularly lamellate, pale to orange in color. Clamp connections are present. The spores are small, hyaline, and cylindric. The genus is often placed in other families, as it appears transitional in its characters. Examine specimens of available species (Plate XIV).

Polyporus — The basidiocarps of *Polyporus* are annual and vary from resupinate-reflexed to applanate, to stipitate. The context may be brown or white. The hymenophore is poroid. Examine the prepared slide showing sections through pores of *Polyporus* sp. (Fig. 216). Look at sporophores of the following species and note the characters indicated.

Polyporus berkeleyi — This species forms large, buff-colored, stipitate sporophores.

Pycnoporus (Polyporus) cinnabarinus — In this species the sporophores are applanate and reddish to cinnamon colored.

Laetiporus (Polyporus) sulphureus — When young the sporophores of this species are pale to bright yellow and they form tiers of brackets on the sides of trees. (Plate XV). This is edible when young.

Coriolus (Polyporus) versicolor — The upper surface of the sporophore has multicolored zones and is velvety (Plate XV). Mount a small portion from the edge of a sporophore in water and carefully dissect it apart. Examine under the microscope and look for generative, binding, and skeletal hyphae (Fig. 217-219).

Poria — In *Poria* the sporophore is always resupinate. The context may be white or dark. Examine specimens of *Poria* spp. (Plate XV).

Poria cocos — This species produces a large underground sclerotium that is white inside, with a dark brown, bark-like outer layer. Called "tuckahoe," it was supposedly eaten by the North American Indians.

Poria incrassata — This is the common house rot fungus (Plate XV). Examine boards showing the brown cubical rot produced by this fungus. Some boards bear sporophores.

Bavendamm Test — The identification of wood rotting fungi is important in the pathology of wood products, but identification is often difficult because most wood rotting fungi do not produce sporophores in culture. Identification consequently must be based on hyphal and cultural characteristics. One useful cultural characteristic is the separation of the white and brown rot fungi. One method of doing this is by means of the Bavendamm Test. The fungus to be tested is grown on malt extract agar, then small blocks of mycelium are transferred to plates of tannic or gallic acid media. Within a few days a brown zone is formed by white rot fungi. Examine the demonstration plates of *Heterobasidion annosum* (= *Fomes annosus*) and *Polyporus tulipiferae* and note the results.

REFERENCES

Coker, W.C., and A.H. Beers. 1951. The Stipitate Hydnums of the Eastern United States. *Univ. North Carolina Press*, Chapel Hill.

Corner, E.J.H. 1950. *A Monograph of Clavaria and Allied Genera.* Oxford University Press, Cambridge.

Corner, E.J.H. 1968. *A Monograph of Thelephora (Basidiomycetes).* Nova Hedwigia Supp. 27: 1-110.

Cunningham, G.H. 1963. *The Thelephoraceae of Australia and New Zealand.* Bull. 145, Dept. Scientific and Industrial Research, Wellington.

Lentz, P.L. 1955. *Stereum* and Allied Genera of Fungi in the Upper Mississippi Valley. *Agr. Monogr.* 24, U.S. Dept. Agr., Washington.

Lowe, J.J. 1957. *Polyporaceae of North America. The Genus Fomes.* State Univ. College Forestry, Syracuse Univ., Syracuse.

Lowe, J.J., and R.L. Gilbertson. 1961. Synopsis of the Polyporaceae of the Southeastern United States. *J. Elisha Mitchell Sci. Soc.* 77: 43-61.

Nobles, M.K. 1965. Identification of cultures of wood-inhabiting Hymenomycetes. *Canad. J. Bot.* 43: 1097-1139.

Overholts, L.O. 1953 *The Polyporaceae of the United States, Alaska, and Canada.* Univ. Michigan Press, Ann Arbor.

Stalpers, J.A. 1978. Identification of wood-inhabiting Aphyllophorales in pure culture. *Stud. Mycol.* 16: 1-248.

PLATE XIV. Mature basidiocarps. Upper row, l-r. *Schizophyllum commune.* × 1. *Hydnum imbricatum.* × 1. *Merulius* sp. × 1. Middle row, l-r. *Clavariadelphus pistillaris.* × 1. *Irpex* sp. × 2. *Dentinum repandum.* × 1. Lower row, l-r. *Hericium coralloides.* × 1. *Sparassis crispa.* × 0.25. *Clavicorona pyxidata.* × 1.

FIG. 210. Section through hymenium of *Exobasidium camelliae*. One basidium has two attached basidiospores. x 680. FIG. 211-212. *Russula emetica*. FIG. 211. Sphaerocysts. x 480. FIG. 212. Cystidium. x 480. FIG. 213. Basidium of *Agaricus campestris* with attached basidiospores. x 1000. FIG. 214. Section through pileus, gills, and stipe of *Coprinus* sp. x 10. FIG. 215. Section through gill trama and hymenium of *Coprinus* sp. x 250. FIG. 216. Section through hymenophore of *Polyporus* sp. showing pores. x 82. FIG. 217-219. *Coriolus versicolor*. FIG. 217. Skeletal hypha. x 600. FIG. 218. Binding hyphae. x 600. FIG. 219. Generative hypha with clamp connection (arrow). x 800. FIG. 220-223. *Spiniger meineckellus* (= *Oedocephalum lineatum*), conidial state of *Heterobasidion annosum*. FIG. 220. Denticles forming on ampulla. x 680. FIG. 221. Young conidia forming. x 680. FIG. 222. Mature conidia on conidiophore. x 680. FIG. 223. Mature conidia. x 680.

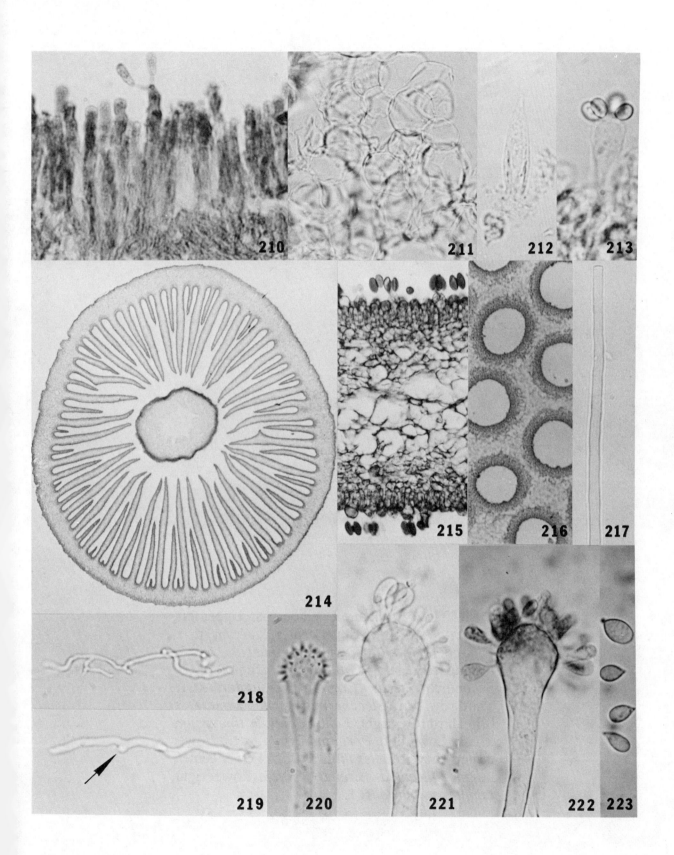

PLATE XV. Mature basidiocarps. Upper left, *Corticium* sp. x 1. Upper center, *Stereum* sp. x 1. Upper right, *Laetiporus sulphureus*. x 0.1. Second row, left. *Ganoderma tsugae*. x 1. Second row, right. *Coriolus versicolor*. x 1. Third row, left. *Steccherinum septentrionale*. x 0.1. Third row, right. *Echinodontium tinctorium*. x 0.3. Lower left, *Poria incrassata*. x 1. Lower center, above. *Fomes pini*. x 0.25. Lower center, below. *Daedalea quercina*. x 0.25. Lower right, *Lenzites saepiaria*. x 1.

FIG. 224. Hypha of *Corticium coeruleum* with clamp connections. × 1500. FIG. 225-226. *Cyathus* sp. FIG. 225. Basidiospores. × 600. FIG. 226. Section through peridiole with hymenium in center. × 63. FIG. 227. Non-mycorrhizal root of loblolly pine. × 4. FIG. 228. Ectomycorrhizae on loblolly pine root formed by *Pisolithus*. × 4. FIG. 229. Longitudinal section through root of loblolly pine showing mycorrhizal mantle and Hartig net. × 63. FIG. 230. Cross section of same. × 63. FIG. 231. Arbuscles of endomycorrhizae in cotton root. × 400. FIG. 232-233. *Physcia* sp. FIG. 232. Section through thallus showing dark cells of algal layer. × 160. FIG. 233. Section through apothecium. × 63. FIG. 227 and 228 courtesy Donald H. Marx, U.S. Forest Service.

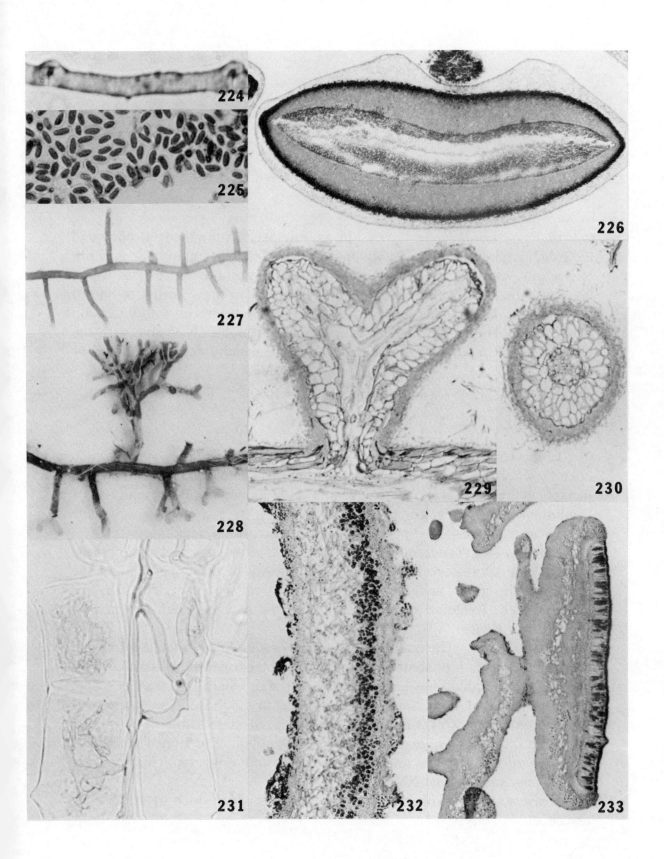

Gasteromycetidae

In the gasteromycetes the basidia are enclosed within the fruit-body and the basidiospores are not forcibly discharged. A hymenium may or may not be present. At maturity the basidiospores form a gleba or spore mass which is usually enclosed within a peridium. The peridium may consist of one or more layers. The hyphae are dolipore septate, and with or without clamp connections. The fruiting bodies range in size from barely visible with the naked eye to the largest of all fungi. They are usually saprobic on soil or wood.

Lycoperdales

In the Lycoperdales the peridium consists of two major layers, which open by a pore or irregular cleavages at maturity. The gleba is powdery and the spores are puffed out of the sporophore. A hymenium is present, as is a capillitium in most species.

Lycoperdaceae — Exoperidium of one layer, often deciduous.

Calvatia — In *Calvatia* the fruit bodies are often large and occur on the ground. The spores are released by falling away of the upper part of the peridium. The spores are globose, and a capillitium is present. The subgleba is chambered. Examine specimens of available species, such as *Calvatia cyathiformis* (Plate XVI).

Lycoperdon — The fruit bodies of *Lycoperdon* are of medium size and grow on the ground, or sometimes on wood. The exoperidium is roughened, and the endoperidium has an apical pore through which the spores escape. There is abundant capillitium, and the spores are globose. The subgleba is chambered (Plate XVI). Examine the prepared slide showing a section through a sporophore of *Lycoperdon* sp. Look at young sectioned specimens of *Lycoperdon pyriforme* and note the gleba and subgleba. Then observe mature specimens (Plate XVII).

Geastraceae — Exoperidium of three layers which open by stellate dehiscence.

Geastrum — In *Geastrum* the endoperidium is papyraceus and persistent, with an apical ostiole. The gleba has a prominent pseudocolumella of highly compacted capillitium. These are the earth stars. Examine specimens of *Geastrum triplex* and note how the exoperidium separates into stellate sections (Plate XVI).

Sclerodermatales

The peridium is thick and tough and usually one-layered. There is no hymenium and the gleba is powdery at maturity.

Sclerodermataceae — Fruit-body simple, peridium not divided into separable layers, capillitium absent.

Scleroderma — The gleba is powdery and continuous at maturity. Dehiscence is irregular and often somewhat stellate. The spores are spiny or reticulate. Examine specimens of *Scleroderma geaster,* the earth ball (Plate XVI). Mount some spores in water and examine under the microscope.

Pisolithus — In *Pisolithus* the gleba is divided into fragile, lenticular peridioles. Dehiscence is by weathering at the apex. The spores are brown, warted, and globose. Examine specimens of *Pisolithus tinctorius* (Plate XVI).

Tulostomatales

The fruit-body is stipitate with a more or less globose fertile portion; there is no hymenium. The gleba is powdery and a capillitium is present.

Calostomataceae — Peridium complex, of four layers.

Calostoma — The peridium has four clearly defined layers. The gleba is borne on a fibrous stipe. Examine specimens of *Calostoma cinnabarina* (Plate XVI).

Nidulariales

In the Nidulariales the fruit-body grows on the ground or on organic debris. The peridium is of one to many layers. The gleba consists of one to many hard peridioles, and the spores are smooth, hyaline, thick-walled, and often large.

Nidulariaceae — Fruit-body with peridium of one to three layers, often forming a cup with several peridioles.

Cyathus — The fruit-bodies grow on the ground, dead wood, or dung. The peridium is five-layered. The peridioles have only a thin tunica, and they are attached by a complicated funicle (Plate XVII). Examine the prepared slide showing sections through a sporophore of *Cyathus* sp. (Fig. 225-226), then examine dried specimens of *Cyathus striatus* (Plate XVII).

Phallales

The phalloid fruit-body is frequently pileate, but without a true stipe. An hymenium is present and the basidia mature more or less simultaneously. At maturity the gleba is fleshy or mucid. The spores are hyaline and bacillar or pale brown and ovoid.

Phallaceae — Receptacle a simple, hollow column bearing the mucid, odiferous gleba near the top, on the outside.

Mutinus — In this genus the receptacle consists of a simple stalk-like column with the gleba borne subapically on the stipe. Examine specimens of *Mutinus caninus* (Plate XVII).

Phallus — In *Phallus* the receptacle has a separable campanulate cap bearing the gleba; the cap covers the apex of the stalk. Examine preserved specimens of *Phallus impudicus* (Plate XVI). Also examine specimens of *Dictyophora duplicata,* which has a net-like indusium under the cap (Plate XVII). These are now included in the genus *Phallus*.

REFERENCES

Brodie, H.J. 1975. *The Bird's Nest Fungi.* Univ. Toronto Press, Toronto.

Coker, W.C., and J.N. Couch. 1928. *The Gasteromycetes of the Eastern United States and Canada.* Univ. North Carolina Press, Chapel Hill.

Smith, A.H., 1951. *Puffballs and their Allies in Michigan.* Univ. Michigan Press, Ann Arbor.

PLATE XVI. Mature basidiocarps. Top, *Geastrum triplex*. Middle left, *Pisolithus tinctorius*. Middle right, *Scleroderma geaster*. Bottom row, l-r. *Calostoma cinnabarina. Calvatia cyathiformis. Phallus impudicus.* All × 1.

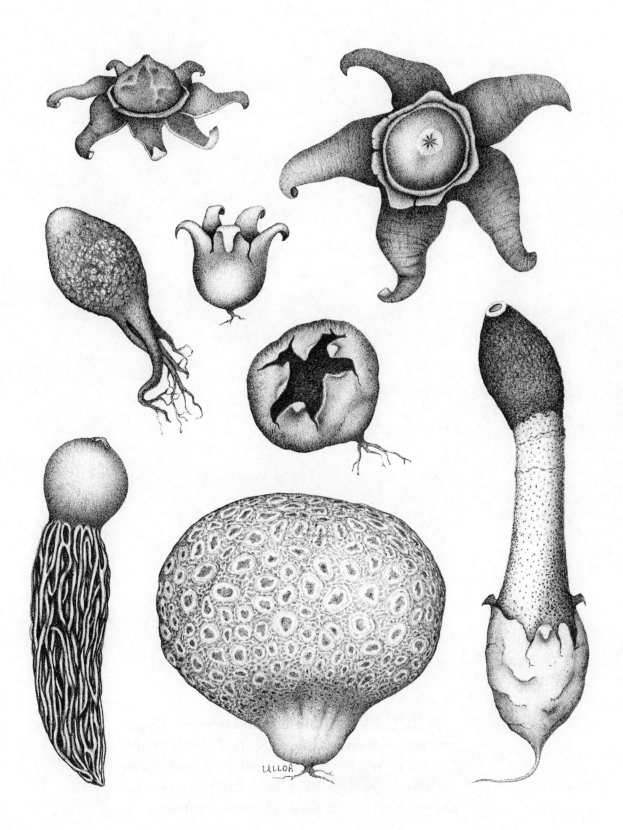

PLATE XVII. Mature basidiocarps. Upper left, *Dictyophora duplicata.* × 0.67. Upper right, *Cyathus striatus.* × 2. Center. *Lycoperdon pyriforme.* × 1. Lower left, *Lycoperdon candidum.* × 1. Lower right, *Mutinus caninus.* × 1.

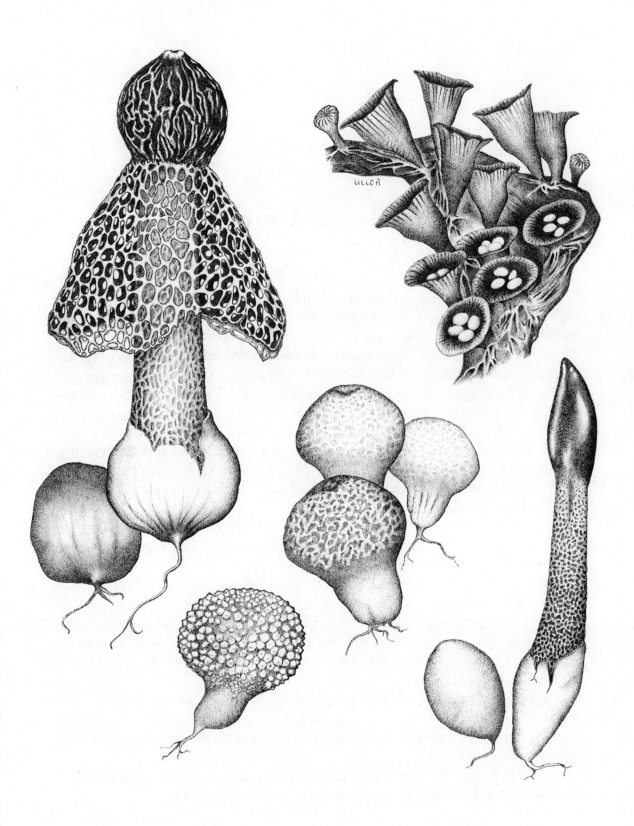

Mycorrhizae

Mycorrhizae are associations formed between fungi and the roots of higher plants. They occur in gymnosperms and in both herbaceous and woody angiosperms. Studies have shown that mycorrhizae greatly benefit the growth of the plant. There are two types of mycorrhizae, the ectomycorrhiza and the endomycorrhiza. Most ectomycorrhizae are formed by agarics and gasteromycetes, whereas endomycorrhizae are largely formed by members of the Endogonaceae.

Ectomycorrhizae form a thick fungus mantle on the outside of the young feeder roots. They are especially common on gymnosperms and they usually cause the roots to be short and much-branched. In addition to the mantle, hyphae grow into the cortical tissues of the root, forming what is known as the Hartig net between the cells.

Look at preserved material of naturally formed ectomycorrhizae on roots of *Pinus taeda,* then look at preserved material of *P. taeda* seedlings with mycorrhizae (Fig. 227-228).

Compare the roots of seedlings of *Pinus elliottii* grown with and without ectomycorrhizae.

Examine prepared slides showing sections through pine roots bearing ectomycorrhizae formed by *Pisolithus tinctorius* and *Thelephora terrestris*. Both cross and longitudinal sections can be found. Note the mantle and the Hartig net (Fig. 229-230).

Endomycorrhizae — In endomycorrhizae the fungus grows mainly inside the root, with hyphal strands growing out into the soil. Endomycorrhizae form swollen intracellular vesicles and arbuscules; most endomycorrhizae are formed by *Gigaspora* and *Glomus*.

Mount a small portion of phloxine-stained, macerated cotton root in water and look for arbuscules and vesicles in the root cells (Fig. 231).

REFERENCES

Brundrett, M.C., Y. Piché, and R.L. Peterson. 1985. A developmental study of the early stages in vesicular-arbuscular mycorrhiza formation. Canad. J. Bot. 63: 184-194.

Bryan, W.C. 1961. Synthetic culture of mycorrhizae of Southern pines. Forest Science 7: 123-129.

Gerdemann, J.W., and J.M. Trappe. 1974. The Endogonaceae in the Pacific Northwest. *Mycol. Mem.* 5: 1-76.

Harley, J.L. 1959. *The Biology of Mycorrhiza.* Leonard Hill, Ltd., London.

Hacskaylo, E. 1971. *Mycorrhizae.* U.S. Govt. Printing Off., Washington, D.C.

Jackson, R.M., and P.A. Mason. 1984. *Mycorrhiza.* Edward Arnold Ltd., London.

Marks, G.C., and T.T. Kozlowski. 1973. *Ectomycorrhizae. Their Ecology and Physiology.* Academic Press, New York.

Marx, D.H., and W.C. Bryan. 1970. Pure culture synthesis of ectomycorrhizae by *Thelephora terrestris* and *Pisolithus tinctorius* on different conifer hosts. *Canad. J. Bot.* 48: 639-643.

Marx, D.H. 1970. The influence of ectotrophic mycorrhizal fungi on the resistance of pine roots to pathogenic infections. V. Resistance of mycorrhizae to infection by vegetative mycelium of *Phytophthora cinnamomi. Phytopathology* 60: 1472-1473.

Miller, S.L., and O.K. Miller, Jr. 1984. Synthesis of *Elaphomyces muricatus + Pinus sylvestris* ectomycorrhizae. *Canad. J. Bot.* 62: 2363-2369.

Pachlewski, R., and J. Pachlewska. 1974. *Studies on symbiotic properties of mycorrhizal fungi of pine (Pinus silvestris L.) with the aid of the method of mycorrhizal synthesis in pure cultures on agar.* Forest Res. Inst., Warsaw.

Powell, C.L., and D.J. Bagyarag. (Eds.). 1984. *VA Mycorrhiza.* CRC Press, Boca Raton.

Sanders, F.E., B. Mosse, and P.B. Tinker. (Eds.). 1974. *Endomycorrhizas.* Academic Press, New York.

Schenck, N.C. (Ed.). 1982. *Methods and Principles of Mycorrhizal Research.* Amer. Phytopathol. Soc., St. Paul.

Trappe, J.M. 1962. Fungus associates of ectotrophic mycorrhizae. *Bot. Rev.* 28: 538-606.

Lichens

Lichens are organisms formed by the association of an alga with a fungus. The lichen thallus differs morphologically from either of the component organisms when they are grown separately. Most of the bulk of the lichen thallus consists of fungus tissue. Since lichens have a photosynthetic component they can grow on barren areas, such as rock outcrops. They are also common on tree bark and soil. In nearly all lichens the fungus is an ascomycete, usually a discomycete. Both unitunicate and bitunicate ascomycetes are involved in lichen formation, and they are classified on this basis, along with other fungi.

Foliose lichens — The thallus is flattened and leaf-like, with the upper and lower surfaces differing in color or surface features.

Hypogymnia — Thallus gray-green above, with long, slender lobes.

Physcia — Examine a prepared slide of a section through a thallus. In this lichen the algal cells are single and somewhat clustered. Also examine a section through an apothecium (Plate XVIII, Fig. 231-233).

Peltigera — Examine a prepared slide of a section through a thallus. Note the relationship of the alga to the fungus, and the filamentous algal thallus.

Peltigera aphthosa — Thallus broadly foliose, light gray-green above when dry, becoming bright green when moistened (Plate XVIII).

Pseudocyphellaria anomala — Thallus large, tan to brown above with ridges on upper surface, dark-brown apothecia common (Plate XVIII).

Umbilicaria — The thallus is leathery, circular, brown and smooth on top, with the lower surface covered with dense, black rhizines. The thallus is umbilicate, i.e., it is attached to the substrate by a single central cord. These are the Rock Tripe lichens.

Fruticose lichens — In fruticose lichens the branches of the thallus are round in cross section, with few differences between top and bottom. They may be bushy, hairy, or strap shaped.

Alectoria fremontii — The thallus is dark brown to almost black, very slender and hair-like.

Cladonia alpestris and *C. rangiferina* — These species form grayish masses of branched colonies; in *C. alpestris* the colonies form discrete, compact heads, whereas in *C. rangiferina* the colonies are much more extensive. *Cladonia rangiferina* is the reindeer moss.

Usnea salacina — This forms long, slender, gray-green thalli on trees (Plate XVIII).

Crustose lichens — The thallus of crustose lichens is appressed to the surface of the substrate, and is sometimes in the substrate, in which case the outer margin is delimited by a dark line or color difference. Some small foliose lichens can be confused with crustose species. These lichens are common on rocks (saxicolous) and tree bark (corticolous).

Cyphelium — This forms gray, superficial thalli on tree bark. Black apothecia are common.

Graphis scripta — The gray thallus in this species forms in the bark of trees and is recognized by the color and dark margin. The apothecia are elongate (Plate XVIII).

Lecanora — This is a large genus of crustose lichens that are common on rock and tree bark. Abundant thalline apothecia are present. The ascospores are one-celled and the paraphyses are unbranched (Plate XVIII).

Squamulose lichens — These lichens are characterized by squamules, small lobe-like structures which lack a lower cortex. They often form the primary thallus of *Cladonia*.

Cladonia chlorophaea — This species forms greenish-gray cups up to 1.5 cm high on coarse squamules (Plate XVIII).

Cladonia cristatella — This is the British Soldier, with yellowish-green podetia that are branched near the tip. The tips commonly bear bright red apothecia (Plate XVIII).

If material is available, make thin sections of lichen apothecia with a razor blade and mount them in water. Examine the sections under the microscope and look for asci and ascospores.

REFERENCES

Ahmadjian, Vernon. 1967. *The Lichen Symbiosis*. Blaisdell Publ. Co., Waltham.
Ahmadjian, V., and M.E. Hale. (Eds.) 1973. *The Lichens*. Academic Press, New York.
Bolton, E.M. 1960. *Lichens for Vegetable Dyeing*. Studio Books, London.
Duncan, Ursula K. 1970. *Introduction to British Lichens*. T. Buncle & Co., Ltd., Arbroath.
Hale, Mason E., Jr. 1961. *Lichen Handbook*. Smithsonian Institution Press, Washington.
Hale, Mason E., Jr. 1969. *How to Know the Lichens*. Wm. C. Brown Co., Dubuque.
Hale, M.E., Jr. 1983. *The Biology of Lichens*. 3rd ed. Edward Arnold, Baltimore.
Hawksworth, D.L., and D.J. Hill. 1984. *The Lichen-Forming Fungi*. Blackie and Son, Ltd., London.
Kershaw, K.A. 1985. *Physiological ecology of lichens*. Cambridge Univ. Press, Cambridge.
Lawrey, J.D. 1984. *Biology of Lichenized Fungi*. Praeger Pub., New York.
Martin, W., and J. Child. 1972. *Lichens of New Zealand*. A.H. & A.W. Reed Ltd., Wellington.
Rogers, R.W. 1981. *The Genera of Australian Lichens (lichenized fungi)*. Univ. Queensland Press, Queensland.
Seaward, M.R.D. (Ed.). 1977. *Lichen Ecology*. Academic Press, London.
Taylor, Conan J. 1967. *The Lichens of Ohio. Part I. Foliose Lichens*. Ohio State Univ., Columbus.

Taylor, Conan J. 1968. *The Lichens of Ohio. Part II. Fructicose and Cladoniform Lichens.* Ohio State Univ., Columbus.

Thomson, J.W. 1967. *The Lichen Genus Cladonia in North America.* Univ. Toronto Press, Toronto.

Thomson, J.W. 1979. *Lichens of the Alaskan Arctic Slope.* Univ. Toronto Press, Toronto.

Thomson, J.W. 1984. *American Arctic Lichens.* 1. *The Macrolichens.* Columbia Univ. Press, New York.

PLATE XVIII. Mature lichen thalli. Upper left, *Lecanora* sp. × 1. Upper right, *Pseudocyphellaria anomala.* x 1. Center, *Usnea* sp. x 1. Middle right, *Graphis scripta.* above, *Peltigera aphthosa.* below. x 1. Lower left, *Parmelia* sp. x 1. Lower center, *Cladonia cristatella.* x. 10. Lower right, *Cladonia chlorophaea. x 2.*

Fossil Fungi

In recent years a considerable number of fossil fungi have been discovered. Some of these are so similar to extant species that they can be assigned to modern genera.

Examine the demonstration slide containing portions of "Sapindus" leaflets of Eocene Age from Henry Co., Tennessee clay pits. Observe the fossil fruiting bodies and hyphae on the leaf surface. The fruiting body resembles some modern ascomycetes.

REFERENCES

Dilcher, D.L. 1963. Eocene epiphyllous fungi. *Science* 142: 667-669.

Dilcher, D.L. 1965. Epiphyllous fungi from Eocene deposits in Western Tennessee, U.S.A. *Palaeontographica* 116: 1-54.

Lange, R.T., and P.H. Smith. 1975. *Ctenosporites* and other Paleogene fungal spores. *Canad. J. Bot.* 53: 1156-1157.

Pirozynski, K.A. 1976. Fossil fungi. *Annu. Rev. Phytopathol.* 14: 237-246.

Sherwood-Pike, M.A., and J. Gray. 1985. Silurian fungal remains: probable records of the class Ascomycetes. *Lethaia* 18: 1-20.

Stubblefield, S.P., C.E. Miller, T.N. Taylor, and G.T. Cole. 1985. *Geotrichites glaesarius*, a conidial fungus form tertiary Domincan amber. *Mycologia* 77: 11-16.

Stubblefield, S.P., and T.N. Taylor. 1983. Studies of Paleozoic fungi. I. The structure and organization of *Traguairia* (Ascomycota). *Amer. J. Bot.* 70: 387-399.

Stubblefield, S.P., and T.N. Taylor. 1983. Studies of Carboniferous fungi. II. The structure and organization of *Mycocarpon, Sporocarpon Dubiocarpon* and *Coleocarpon* (Ascomycotina). *Amer. J. Bot.* 70: 1482-1498.

Stubblefield, S.P., T.N. Taylor, and C.B. Beck. 1985. Studies of Palezoic fungi. IV. Wood-decaying fungi in *Callixylon newberryi* from the Upper Devonian. *Amer. J. Bot.* 72: 1765-1774.

Wolf, F.A. 1968. Fungus spores in Lake Singletary sediment. *J. Elisha Mitchell Sci. Soc.* 84: 227-232.

Fungal Ecology

The ecology of fungi is the study of fungi in relation to their environment and to each other. One aspect of fungal ecology is the determination of which species make up particular communities, and how these may be associated with the higher plants in the community. A second area of study involves defining the role of fungi in a given community. Many of these studies are of practical significance; our knowledge of the role mycorrhizal fungi play in the growth of pine trees makes it possible to grow them in areas where formerly they would not grow. Many studies in plant pathology that deal with the host-parasite relationship and possible disease control measures are basically ecological in nature. Thus studies on the ecology of fungi will continue to be important in the future. The same is true for the medical and veterinary mycology fields.

Fungal Succession — The succession of fungi that occurs on a given substrate can be followed readily in the laboratory. One convenient method of doing this is through observation of coprophilous fungi. This can be done by collecting some herbivore dung and placing it on filter paper in a damp chamber. Rabbit dung works well. Incubate the dung for about three days, then make daily observations and record the fungi that appear. Continue this for at least three weeks. Upon completion of the study, compile a list of the fungi encountered, when they occurred, and how long they were present. Your list should reveal a definite succession of fungi of different higher taxa during the observation period, starting with phycomycetes, which are followed by ascomycetes and basidiomycetes.

The dung used to study *Pilobolus* can be retained as a class demonstration in fungal succession.

Other substrates, such as leaf litter, can also be used in succession studies.

Species Diversity and Community Make-Up. Although some fungi are restricted in their host or substrate range, many species occur in a wide range of habitats. Species diversity and community populations can be studied using soils and the dilution plate technique. Collect soil samples and plate them out according to the instructions. Examine the plates frequently and record number of colonies per plate and the number of each species observed. If the soil samples are taken from very different habitats, such as a forest floor and a beach, you can compare both the species composition and the population of the different communities.

A similar study can be conducted on seeds of various kinds. To determine the fungal flora of seeds it is necessary to remove surface contaminants before plating them on agar. To do this, immerse the seeds for 2-3 minutes in a solution of 20 ml bleach: 80 ml water, then remove with sterile forceps and allow excess fluid to drain off. Place the seeds on a suitable medium, such as rose bengal medium (RBM-2) (Fig. 234). Some seeds should also be plated on malt-salt (MSA) or Harrold's Agar (M4OY) to detect osmophilic fungi. Depending upon the size of the seeds, several can be placed in each petri dish. For larger seeds 20 mm deep dishes should be used. Incubate at room temperature for 4-5 days and observe fungi as they appear (Fig. 235). For proper species identification it is usually necessary to subculture individual colonies in test tubes so they do not become contaminated.

Role of Fungi. Fungi are basically decomposers which release nutrients that are utilized by themselves and by other organisms. When this decomposer occurs on items we use, such as food or wood products, the process becomes economically important.

Fungal decomposition of fruits can be demonstrated readily in the laboratory by inoculating oranges or other citrus fruits with spores of *Penicillium italicum*. Make a small wound through the peel and introduce spores from a culture of the fungus. Place the inoculated fruit in a damp chamber or plastic bag and observe the results after a few days.

Other organisms that respond well are *Penicillium expansum* on apple (Fig. 236-237) or *Monilia fructicola* on peach.

Litter decomposition can also be demonstrated, but this takes longer. Collect litter, such as leaves from a forest floor, and divide it into 10-12 lots of equal weight. Hay can also be used. Record the initial wet and dry weights of a portion of the sample. Place the lots in aluminum foil or other suitable container and leave them at room temperature, being certain the litter has ample aeration. Keep the litter moist; if you wish to do a critical study you should measure the amount of water added and add the same amount each time. Each week, examine two of the litter samples for fungi, using the dissecting and light microscope. Dry the lots, weigh, and record the loss in weight. After 10-12 weeks considerable decomposition should have occurred. Tabulate the results for the class.

Litter decomposition in large piles often involves heat accumulation (self-heating) and the growth of thermophilic fungi. To isolate these fungi, materials must be incubated at 45-50C. Two broad groups of fungi will be recovered, thermotolerant species that will also grow at room temperature, and the obligate thermophiles that must be grown at 45-50C for proper growth and sporutation. Note the differences in number of species and species composition from materials incubated at 20-25C and 45-50C.

Decomposition can also be demonstrated with wood blocks and a woodrot fungus.

One role of fungi in soil can be demonstrated with nematode trapping species such as *Arthrobotrys dactyloides*. This fungus forms constricting rings on the hyphae (Fig. 239). When a nematode enters a ring (Fig. 238) it constricts around the body of the nematode (Fig. 240), holding it firmly and allowing the hyphae to colonize the body of the nematode. Examine petri plates containing *Arthrobotrys dactyloides* and nematodes and look for rings and trapped nematodes. These can be best observed under medium power of the compound microscope. You can also remove some of the fungus-nematode culture and mount it on a slide in a drop of water, add a cover slip, and examine it under the microscope.

REFERENCES

Badura, L. 1965. Investigations on the soil mycoflora of a beech community in the Botanical Garden of the Turin University (Italy). *Fragmenta Floristics et Geobotanical* 11: 197-2080.

Badurowa, M., and L. Badura. 1967. Further investigations on the relationship between soil fungi and the macroflora. *Acta. Soc. Bot. Pol.* 36: 515-529.

Barron, G.L. 1977. The Nematode-Destroying Fungi. Canad. Biol. Pub., Guelph.

Brown, J.C. 1958. Soil fungi of some British sand dunes in relation to soil type and succession. *J. Ecology* 46: 641-664.

Christensen, M. 1969. Soil microfungi of dry to mesic conifer-hardwood forests in northern Wisconsin. *Ecology* 50: 9-27.

Clark, F.E. 1970. *Decomposition of organic materials in grassland soil.* U.S. IBP Grassland Biome, Technical Report No. 61.

Clark, F.E., and E.A. Paul. 1970. The microflora of grassland. *Advan. Agronomy* 22: 375-435.

Clarke, D.C., and M. Christensen. 1981. The soil microfungal community of a South Dakota grassland. *Canad. J. Bot.* 59: 1950-1960.

Cooney, D.G., and R. Emerson. 1964. *Thermophilic Fungi.* W.H. Freeman and Co., San Francisco.

Flannigan, B., and C.S. Sagoo. 1977. Degradation of wood by *Aspergillus fumigatus* isolated from self-heated wood chips. *Mycologia* 69: 514-523.

Gochenaur, S.E., and W.F. Whittingham. 1967. Mycoecology of willow and cottonwood lowland communities in southern Wisconsin. I. Soil microfungi in the willow-cottonwood forests. *Mycopathol. Mycol. Appl.* 33: 125-139.

Griffin, D.M. 1972. *Ecology of Soil Fungi.* Syracuse University Press, Syracuse.

Harley, J.L. 1971. Fungi in ecosystems. *J. Ecology.* 59: 653-668.

Hoover-Litty, H., and R.T. Hanlin. 1985. The mycoflora of wood chips to be used as mulch. *Mycologia* 77: 721-731.

Hudson, H.J. 1968. The ecology of fungi on plant remains above the soil. *New Phytologist* 67: 837-874.

Kuthubutheen, A.J., and J. Webster. 1986. Water availability and the coprophilous fungus succession. *Trans. Brit. Mycol. Soc.* 86: 63-76.

Larsen, K. 1971. Danish endocoprophilous fungi and their sequence of occurrence. *Botanisk Tidsskrift* 66: 1-32.

Ofosu-Asiedu, A., and R.S. Smith. 1973. Degradation of three softwoods by thermophilic and thermotolerant fungi. *Mycologia* 65: 240-244.

Parkinson, D., and J.S. Waid. (Eds.). 1960. The Ecology of Soil Fungi. Liverpool Univ. Press, Liverpool.

Richardson, M.J. 1972. Coprophilous ascomycetes on different dung types. *Trans. Brit. Mycol. Soc.* 58: 37-48.

Rossi L., E.A. Fano, A. Basset, C. Fanelli, and A.A. Fabbri. 1983. An experimental study of a microfungal community on plant detritis in a Mediterranean woodland stream. *Mycologia* 75: 887-896.

Sewell, G.W.F. 1959. The ecology of fungi in *Calluna*-healthland soils. *New Phytologist* 58: 5-15.

Smith, R.S., and A. Ofosu-Asiedu. 1972. Distribution of thermophilic and thermotolerant fungi in a spruce-pine chip pile. *Canad. J. Forest. Res.* 2: 16-26.

Tansey, M.R. 1971. Isolation of thermophilic fungi from self-heated, industrial wood chip piles. *Mycologia* 63: 537-547.

Thornton, R.H. 1956. Fungi occurring in mixed oakwood and heath soil profiles. *Trans. Brit. Mycol. Soc.* 39: 485-494.

Visser, S., and D. Parkinson. 1975. Fungal succession on aspen poplar leaf litter. *Canad. J. Bot.* 53: 1640-1651.

Wicklow, D.T., and S.K. Angel. 1983. Some reproductive characteristics of coprophilous ascomycetes in three prairie ecosystems. *Mycologia* 75: 1070-1073.

Wicklow, D.T., and G.C. Carroll. (Eds.) 1981. *The Fungal Community. Its Organization and Role in the Ecosystem.* Marcel Dekker, Inc., New York.

Wicklow, D.T., and V. Moore. 1974. Effect of incubation temperature on the coprophilous fungal succession. *Trans. Brit. Mycol. Soc.* 62: 411-415.

FIG. 234. Peanut seed plated on agar. x 0.73. FIG. 235. Fungal colonies on surface-disinfested peanut seed after 10 days incubation at room temperature. x 0.77. FIG. 236. Lesion produced on apple cv. Golden Delicious 10 days after inoculation with *Penicillium expansum*. x 0.71. FIG. 237. Internal rot of apple tissue caused by *Penicillium expansum*. x 0.71. FIG. 238. Nematode inside unconstricted ring (arrow) of *Arthrobotrys dactyloides*. x 1200. FIG. 239. Two open rings of *Arthrobotrys dactyloides* on hypha. Insert: Constricted ring; the 3 cells comprising the ring are visible. x 1200. FIG. 240. Nematode trapped in two rings (arrows) of *Arthryobotrys dactyloides*. x 500. FIG. 238-240 courtesy Patricia Lappe, Univsersity of Mexico.

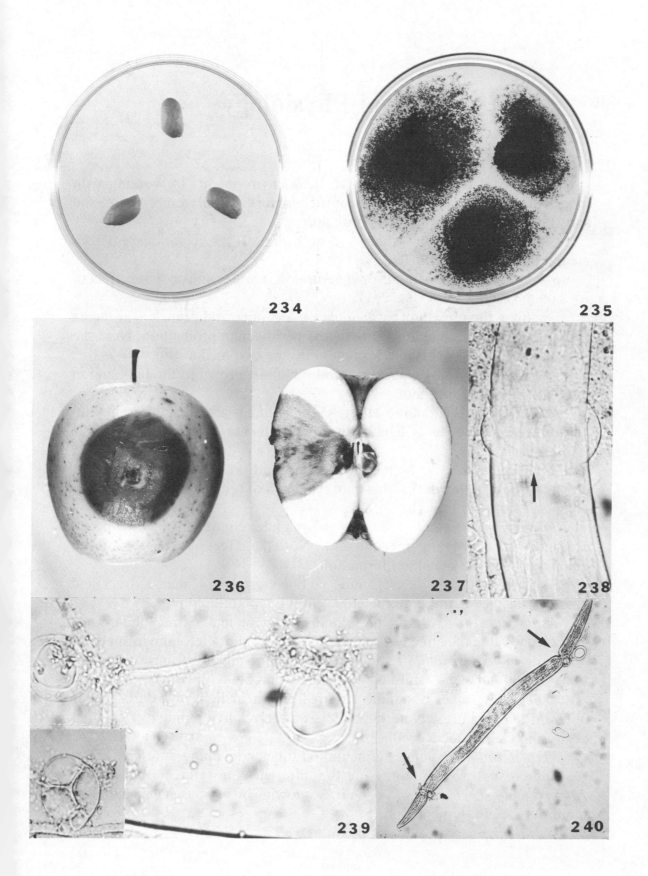

Fungal Physiology

Fungi react in different ways to their environment. In working with fungi it is useful to know the effects of such factors as light, temperature, and substrate on their growth and sporulation.

Observe the following fungi and note the effects of various environmental factors on the fungus.

Sordaria fimicola. Mount some perithecia in water and note the effect of light on the ostiolar neck (Fig. 241).

Compare the growth of *S. fimicola* on V-8 juice and czapek agar media and note the effect on colony growth and perithecium production. Note the time required for perithecium formation on the two media (Fig. 242-244).

Helminthosporium rostratum. Mount conidia grown in the light and in the dark and compare them.

Mucor pusillus. Examine cultures of this fungus grown at different temperatures. On the basis of temperature requirement, what kind of fungus is *M. pusillus*?

Phycomyces nitens. Note the effect of light on the growth of sporangiophores in light and dark. Also note the effect of red and blue wavelengths on sporangiophore growth.

Trichoderma. Examine colonies grown under continuous light, continuous dark, and alternating light and dark. Note the effect of these regimes on colony growth and conidium formation (Fig. 245-246).

Aspergillus versicolor. Examine colonies grown on malt extract and czapek agar media and note the effect on the type of colony.

Aspergillus flavus. Examine colonies grown on malt extract and czapek agar media. Note the effect on colony type, color, and sclerotium formation.

Aspergillus restrictus. Examine colonies growing on various concentrations of salt and sugar, and note the effects on colony growth.

Aspergillus ruber. Examine cultures grown in light and dark and at different temperatures. Note the effects on the relative number of conidial heads and cleistothecia produced.

Fermentation. Examine fermentation tubes containing different yeasts growing on sucrose solution. Note the differences in fermentation ability. Also examine *Saccharomyces cerevisiae* growing on various sugars and note the difference in fermentation.

REFERENCES

Barnett, H.L. 1985. *Fungus Physiology Research at the West Virginia Agricultural and Forestry Experiment Station 1922-1982.* W.Va. Agr. Forest. Expt. Sta. Misc. Pub. 13: 1-113.

Cochrane, V.C. 1958. *Physiology of Fungi.* John Wiley & Sons, Inc., New York.

Griffin, D.H. 1981. *Fungal Physiology.* John Wiley & Sons, New York.

Hawker, L.E. 1957. *The Physiology of Reproduction in Fungi.* Cambridge Univ. Press, London.

Lilly, V.G., and H.L. Barnett. 1951. Physiology of the Fungi. McGraw-Hill Book Co., Inc., New York.

Smith, J.E., and D.R. Berry. (Eds.). 1976. *The Filamentous Fungi.* Vol. 2. *Biosynthesis and Metabolism.* Edward Arnold, London.

Smith, J.E., and D.R. Berry. 1978. *The Filamentous Fungi.* Vol. 3. Developmental Mycology. John Wiley & Sons, New York.

The Physiology of Fungi and Fungus Diseases. Bulletin 488T, W.Va. Univ. Agr. Exp. Sta., Morgantown. 1963.

FIG. 241. Perithecium of *Sordaria fimicola* with ostiolar neck curved in response to side lighting. Compare with FIG. 143. x 112. 242-244. Effect of medium on colony formation in *Sordaria fimicola* after 10 days. FIG. 242. Growth on V-8 juice agar. Black structures are perithecia. x 51. FIG. 243. Growth on malt extract agar. Note difference in texture and lack of perithecia. x 51. FIG. 244. Growth on Czapek agar. Note sparse mycelium and lack of perithecia. x 51. FIG. 245. Colony of *Trichoderma* grown in alternating light and dark. x 0.71. FIG. 246. Growth of *Trichoderma* under continuous light. x 0.71.

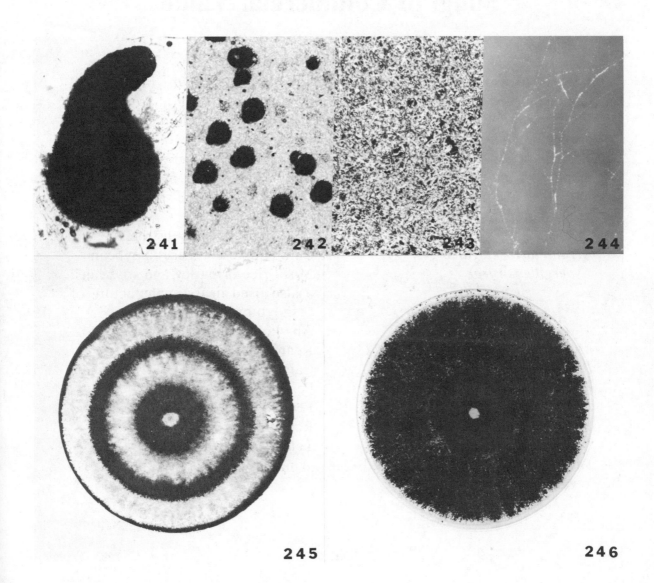

Fungi of Commercial Value

The uses of fungi in industry are many, ranging from the brewing and baking industries to their use in penicillin production and food preparation. Fermented foods are especially common in Asia, although fermented beverages and cheese are used in many parts of the world.

Below are listed several species of fungi that are used commercially, along with the product and usual substrate. Some species may be used in more than one product, and in some instances more than one species may be used to prepare a particular product.

Fungus	Product and substrate
Actinomucor elegans	Sufu (Chinese cheese); soybean
Aspergillus oryzae	Miso; rice or barley and soybeans; also used in the manufacture of shoyu or soy sauce.
Monascus purpureus	Ang-kok (Chinese red rice); rice
Mucor miehie	Rennilase
Penicillium camemberti	Camembert cheese
Penicillium chrysogenum	Penicillin
Penicillium notatum	Penicillin
Penicillium roqueforti	Roquefort cheese
Rhizopus arrhizus	Tempeh; soybean
Saccharomyces cerevisiae var. *ellipsoideus*	Wine; grape and other fruit juices

REFERENCES

Christensen, C.M. 1965. *The Molds and Man*. 3rd Ed. Univ. Minnesota Press, Minneapolis.

Gray, W.D. 1959. *The Relation of Fungi to Human Affairs*. Holt and Co., New York.

Gray, W.D. 1970. *The Use of Fungi as Food and in Food Processing. Pt. 1.* CRC Press, Cleveland.

Gray, W.D. 1973. *The Use of Fungi as Food and in Food Processing, Pt. 2.* CRC Press, Cleveland.

Hesseltine, C.W. 1985. Fungi, people, and soybeans. *Mycologia* 77: 505-525.

Kavalier, L. 1965. *Mushrooms, Molds and Miracles; the strange realm of fungi.* John Day Co., New York.

Smith, G. 1969. *An Introduction to Industrial Mycology*, 6th Ed. St. Martin's Press, New York.

Smith, J.E., and D.R. Berry. (eds.). 1975. *The Filamentous Fungi.* Vol. 1. *Industrial Mycology.* Edward Arnold, London.

Mycotoxins

The word mycotoxin, as it is commonly used, refers to fungal metabolites that are toxic to animals, including man. Mycotoxins affect animals in different ways, from weight loss, tremors, and loss of muscle coordination to causing cancer and death of the animal. These compounds are usually found in foods or feeds and are ingested by the animal while eating. As a group, mycotoxic compounds are relatively heat stable, they are not antigenic, and they are effective in very small amounts. Mycotoxin production usually varies from strain to strain among mycotoxigenic species of fungi.

Below are listed some common toxigenic fungi and their primary metabolite. Where available, examine cultures of these fungi.

An asterisk (*) indicates a carcinogenic compound. Some fungi produce more than one toxin.

Fungus	Primary toxic metabolite
Amanita muscaria	Muscimol
Amanita phalloides	Amanitin
Aspergillus flavus	Aflatoxin*
Aspergillus parasiticus	Aflatoxin*
Aspergillus ochraceus	Ochratoxin
Aspergillus versicolor	Sterigmatocystin*
Aspergillus clavatus	Patulin*
Aspergillus terreus	Citrinin
Aspergillus tamarii	Kojic acid
Balansia epichloe	Agroclavine, elymoclavine
Claviceps purpurea	Ergometrine, ergotamine, and ergotoxines.
Fusarium graminearum	Zearalenone
Fusarium nivale	Nivalenol
Penicillium citrinum	Citrinin
Penicillium crustosum	Penitren A
Penicillium cyclopium	Penicillic acid and Cyclopiazonic acid
Penicillium expansum	Patulin*
Penicillium islandicum	Islanditoxin and Luteoskyrin
Penicillium purpurogenum	Rubratoxin
Phomopsis sp.	Cytochalosin H
Sclerotinia sclerotiorum	Bergapten and Xanthotoxin

Examine vials of aflatoxin dissolved in chloroform and note the differences in the intensity of the yellow pigment. The pigment is not related to aflatoxin but its intensity is correlated with the amount of aflatoxin in solution.

Examine thin layer chromatograph (TLC) plate under UV light and note blue and green spots of standard and sample. Spots in order from top in standard are: B1, B2, G1, and G2.

REFERENCES

Bacon, C.W., J.K. Porter, and J.D. Robbins. 1979. Laboratory production of ergot alkaloids by species of *Balansia*. *J. Gen. Microbiol.* 113: 119-126.

Bacon, C.W., J.K. Porter, and J.D. Robbins. 1981. Ergot alkaloid biosynthesis by isolates of *Balansia epichloe* and *Balansia henningsiana*. *Canad. J. Bot.* 59: 2534-2538.

Berde, B., and H.O. Schild. (eds.). 1978. *Ergot Alkaloids and Related Compounds*. Springer-Verlag, Berlin.

Brook, P.J., and E.P. White. 1966. Fungus toxins affecting mammals. *Ann. Rev. Phytopathol.* 4: 171-194.

Christensen, C.M. 1975. *Molds, Mushrooms, and Mycotoxins*. Univ. Minnesota Press, Minneapolis.

Ciegler, A., S. Kadis, and S.J. Ajl. (eds.). 1971. *Microbial Toxins. Vol. VI. Fungal Toxins*. Academic Press, New York.

Cole, R.J., and R.H. Cox. 1981. *Handbook of Toxic Fungal Metabolites*. Academic Press, New York.

Goldblatt, L.A. (ed.). 1969. *Aflatoxin*. Academic Press, New York.

Herzberg, M. (ed.). 1970. *Toxic Micro-Organisms*, U.S. Govt. Print. Off., Washington, D.C.

Hesseltine, C.W. 1974. Natural occurrence of mycotoxins in cereals. *Mycopath. Mycol. Appl.* 53: 141-153.

Hinton, D.M., and C.W. Bacon. 1985. The distribution and ultrastructure of the endophyte of toxic tall fescue. *Canad. J. Bot.* 63: 36-42.

Kadis, S., A. Ciegler, and S.J. Ajl. (eds.). 1971. *Microbial Toxins. Vol. VII. Algal and Fungal Toxins*. Academic Press, New York.

Kadis, S., A. Ciegler, and S.J. Ajl. (eds.). 1972. *Microbial Toxins. Vol. VIII. Fungal Toxins*. Academic Press, New York.

Lincoff, G., and D.H. Mitchel. 1977. *Toxic and Hallucinogenic Mushroom Poisoning*. Van Nostrand Reinhold Co., New York.

Marasas, W.F.O., P.E. Nelson, and T.A. Tousson. 1984. *Toxigenic Fusarium Species. Identity and Mycotoxicology*. Pennsylvania State Univ. Press, University Park.

Mirocha, C.J., and C.M. Christensen. 1974. Fungus metabolites toxic to animals. *Ann. Rev. Phytopath.* 12: 303-330.

Rodricks, J.V., C.W. Hesseltine, and M.A. Mehlman. 1977. *Mycotoxins in Human and Animal Health*. Pathotox. Publ., Inc., Park Forest South.

Wogan, G.N. (ed.). 1965. *Mycotoxins in Foodstuffs*. M.I.T. Press, Cambridge, Mass.

Wray, B.B., and K.G. O'Steen. 1975. Mycotoxin-producing fungi from house associated with leukemia. *Arch. Environ. Health* 30: 571-573.

INDEX TO GENERA AND SPECIES

Names preceded by an asterisk (*) are illustrated on the page number in italics.

Achlya, 44
*Achlya americana, 44, 46 *47*
*Actinomucor elegans, 188
Agaricus, 144
Agaricus bisporus, 144
Agaricus brunnescens, 144
*Agaricus campestris, 144, 152, *153*, 160, *161*
Agaricus campestris var. *bisporus*, 144
Albugo, 45
*Albugo bliti, 45, 46, *47*
*Albugo candida 45, 46, *47*
Alectoria fremontii, 174
Aleuria, 113
*Aleuria aurantia, 113, 114, *115*, 116, *117*
Allochytridium, 35
Allomyces, 16, 36
*Allomyces javanicus, 36, 38, *39*, 40, *41*, 46, 47
*Alternaria, 64
*Alternaria alternata, 64, 68, *69*
Amanita, 145
Amanita caesarea, 145
*Amanita muscaria, 145, 148, *149*, 189
Amanita phalloides, 189
*Amanita umbonata, *145*, 148, *149*
Amanita verna, 145
*Amanita virosa, 145, 148, *149*
Apiosporina, 131
*Apiosporina morbosa, 131, 134, *135*
*Arcyria, 28, 32, *33*
*Arcyria incarnata, 30, *31*
Armillaria mellea, 146
Armillariella, 146
*Armillariella mellea, 146, 152, *153*
Armillariella tabescens, 146
*Arthrobotrys dactyloides, 181, 182, *183*
Aspergillus, 12, 13, 63, 80
*Aspergillus clavatus, 63, 68, *69*, 189
*Aspergillus flavus, 63, 68, *69*, 184, 189
Aspergillus niger, 63
Aspergillus ochraceus, 189
Aspergillus oryzae, 188
Aspergillus parasiticus, 189
Aspergillus restrictus, 184
Aspergillus ruber, 184
Aspergillus rugulosus, 80
Aspergillus tamarii, 189
Aspergillus terreus, 189

Aspergillus versicolor, 184, 189
Auricularia, 141
*Auricularia auricula, 138, *139*, 141

Balansia, 108
Balansia epichloe, 189
Bipolaris, 64
*Bipolaris maydis, 64, 70, *71*
Blakeslea, 50
*Blakeslea trispora, 50, 58, *59*
*Boletus, 144
*Boletus edulis, 144, 152, *153*

Calocera, 142
*Calocera viscosa, 138, *139*, 142
Calostoma, 167
*Calostoma cinnabarina, 167, 168, *169*
Calvatia, 166
*Calvatia cyathiformis, 166, 168, *169*
Calycella, 119
*Calycella citrina, 119, 120, *121*
Camellia, 143
*Candida albicans, 70, *71*, 72
Cantharellus, 145
*Cantharellus cibarius, 145, 152, *153*
Cephaloascus, 74
Cephaloascus albidus, 74
*Ceratiomyxa fruticulosa, 27, 30, *31*
Ceratocystis, 81
*Ceratocystis fimbriata, 76, *77*, 81
Ceratocystis minor, 81
Ceratocystis pilifera, 81
Ceratocystis ulmi, 76, 81
Chaetomium, 88
*Chaetomium cochliodes, 88, 100, *101*
*Chaetomium funicolum, 88, 100, *101*
*Chaetomium globosum, 88, 90, *91*
Chalara elegans, 64, 149
Chlorophyllum, 145
*Chlorophyllum molybdites, 145, 150, *151*
Choanephora, 50
*Choanephora cucurbitarum, 50, 58, *59*
*Chromelosporium ollare, 113, 116, *117*
Cladonia alpestris, 174
*Cladonia chlorophaea, 175, 176, *177*
*Cladonia cristatella, 175, 176, *177*
Cladonia rangiferina, 174

Cladosporium morbosum, 131
Clavaria, 154
Clavariadelphus, 154
*Clavariadelphus pistillaris, 154, 158, *159*
Claviceps, 108
*Claviceps gigantea, 110, *111*
*Claviceps purpurea, 104, *105*, 108, 110, *111*, 189
Clavicorona, 154
*Clavicorona pyxidata, 154, 158, *159*
Clitocybe, 146
*Clitocybe dealbata, 146, 152, *153*
Clitocybe tabescens, 146
Coemansia, 51
*Coemansia reversa, 58, *59*
Coleosporium, 137
*Coleosporium solidaginis, 134, *135*, 137
Colletotrichum, 65, 89
*Colletotrichum circinans, 65, 70, *71*
Comatricha, 28
*Coprinus, 145, 160, *161*
Coprinus comatus, 145
*Coprinus micaceus, 145, 150, *151*
Cordyceps, 108
*Cordyceps capitata, 108, 110, *111*
Cordyceps clavulata, 108
*Cordyceps melolonthae var. rickii, 110, *111*
*Coriolus versicolor, 157, 160, *161*, 162, *163*
*Corticium, 155, 162, *163*
*Corticium coeruleum, 155, 164, *165*
Craterellus, 145
*Craterellus cantharellus, 145
*Craterellus cornucopioides, 152, *153*
Cronartium, 137
*Cronartium fusiforme, 134, *135*, 137
Cronartium quercuum f. sp. fusiforme, 137
Cryphonectria, 102
*Cryphonectria parasitica, 102, 104, *105*
Cunninghamella, 50
*Cunninghamella echinulata, 50, 58, *59*
Curvularia, 64
*Curvularia geniculata, 68, *69*
*Cyathus, 164, *165*, 167
*Cyathus striatus, 167, 170, *171*
*Cylindrochytridium johnstonii, 35, 38, *39*
Cyphelium, 175
Cystopus candidus, 45

Dacrymyces, 142
Dacrymyces deliquescens, 142
*Dacrymyces palmatus, 138, *139*, 142
Daedalea, 155

Daedalea ambiqgua, 155
*Daedalea quercina, 155, 162, *163*
*Dendrostibella, 64, 70, *71*
Dentinum, 155
*Dentinum repandum, 155, 158, *159*
Diachea, 28
*Diachea leucopodia, 30, *31*
Diaporthe, 102
Diaporthe phaseolorum, 102
Dibotryon morbosum, 131
Dictydium, 29
*Dictydium cancellatum, 29, 30, *31*
*Dictyophora duplicata, 168, 170, *171*
Dictyostelim, 23
*Dictyostelium discoideum, 23, *25,26*
Dipodascopsis, 74
Dipodascopsis uninucleatus, 74, 76, 77
Dipodascus, 74
Dipodascus uninucleatus, 74
Dothidella ulmea, 102

Echinodotium, 155
*Echinodontium tinctorium, 155, 162, *163*
Elaphomyces, 108, 110, *111*
Elsinoe, 127
Elsinoe corni, 127
Emericella, 80
*Emericella rugulosa, 76, *77*, 80
Empusa muscae, 51
Endothia, 102
*Endothia gyrosa, 102, 104, *105*
Endothia parasitica, 102
Endothiella, 102
Entomophthora, 51
*Entomophthora muscae, 51, 60, *61*
Epicoccum, 64
*Epicoccum nigrum, 64, 68, *69*
*Epidermophyton floccosum, 70, *71*, 72
Erysiphe, 84
*Erysiphe graminis, 84, 86, *87*,
Escherichia coli, 3
*Eurotium, 76, *77*, 80
Exidia, 140
*Exidia glandulosa, 138, *139*, 140
Exobasidium, 143
Exobasidium azaleae, 143
*Exobasidium cammelliae, 143, 160, *161*

Fomes, 156
Fomes annosus, 156, 157
*Fomes pini, 156, 162, *163*
Fuligo, 28

*Fuligo cinerea, 28, 30, *31*
Fusarium graminearum, 189
Fusarium nivale, 189
*Fusarium solani, 104, *105*, 106

Ganoderma, 156
Ganoderma curtisii, 156
*Ganoderma tsugae, 156, 162, *163*
Geastrum, 167
*Geastrum triplex, 167, 168, *169*
*Geoglossum, 116, *117*, 118
*Geoglossum difforme, 116, *117*, 119, 120, *121*
Geotrichum, 63
*Geotrichum candidum, 68, *69*
Gigaspora, 172
Gliocladium, 63
*Gliocladium roseum, 63, 68, *69*
Glomerella, 89
*Glomerella cingulata, 104, *105*
Glomus, 172
Gnomonia ulmea, 102
*Graphis scripta, 175, 176, *177*
Graphium ulmi, 81
Gymnoascus, 80
*Gymnoascus reesii, 76, *77*, 80
Gymnosporangium, 137
*Gymnosporangium juniperi-virginianae, 134, *135*, 137
Gymnosporangium nidus-avis, 137
Gyromitra, 113
*Gyromitra esculenta, 113, 114, *115*

Helicosporium, 64
*Helicosporium linderi, 70, *71*
Helminthosporium, 64
Helminthosporium rostratum, 184
Helotium, 119
Helvella, 113
*Helvella lacunosa, 113, 114, *115*
Hemtrichia, 28
*Hemitrichia serpula, 32, *33*
*Hemitrichia stipitata, 30, *31*, 32, *33*
Hericium, 155
*Hericium coralloides, 155, 158, *159*
Hericium erinaceus, 155
Heterobasidion, 155
*Heterobasidion annosum, 156, 157, 160, *161*
Histoplasm capsulatum, 72
Hydnum, 155
*Hydnum imbricatum, 155, 158, *159*
Hypocrea, 106
*Hypocrea rufa, 100, *101*

Hypoderma, 122
Hypogymnia, 174
Hypomyces, 106
*Hypomyces lactifluorum, 104, *105*, 106
*Hypoxylon, 96, *97*, 98
*Hypoxylon multiforme, 98, 100, *101*
*Hysterographium, 133, 134, *135*

*Irpex, 156, 158, *159*
Irpex cinnamomeus, 156

Laboulbenia, 124
*Laboulbenia elongata, 116, *117*, 124
Lactarius, 106, 144
*Lactarius piperatus, 150, *151*
*Laetiporus sulphureus, 156, 162, *163*
*Lecanora, 175, 176, *177*
Lenzites, 156
*Lenzites saepiaria, 156, 162, *163*
Leotia, 119
*Leotia lubrica, 119, 120, *121*
Lepiota, 145
*Lepiota procera, 145, 150, *151*
*Leptosphaeria, 128, *129*, 130
Leptosphaerulina, 132
*Leptosphaerulina crassiasca, 132, 134, *135*
Liquidambar styraciflua, 4
*Lycogala epidendrum, 29, 30, *31*
Lycoperdon, 166
*Lycoperdon candidum, 170, *171*
*Lycoperdon pyriforme, 166, 170, *171*

Marasmius, 146
*Marasmius siccus, 150, *151*
Melanospora, 94
*Melanospora zamiae, 90, *91*, 94, 100, *101*
*Merulius, 156, 158, *159*
*Microsphaera, 84, 90, *91*
*Microsporum gypseum, 70, *71*, 72
Monascus purpureus, 188
Monilia, 118
*Monilia fruticola, 116, *117*, 118, 180
Monilinia, 118
*Monilinia fructicola, 116, *117*, 118, 120, *121*
*Monoblepharis, 36, 46, *47*
*Morchella, 113, 116, *117*
*Morchella esculenta, 113, 114, *115*
Mucor, 49, 50
*Mucor hiemalis, 49, 56, *57*
Mucor miehei, 188
Mucor pusillus, 184
*Mucor rouxianus, 54, *55*

Mutinus, 168
*Mutinus caninus, 168, 170, *171*
Mycosphaerella, 132
*Myriangium, 127, 128, *129*
Myrothecium, 64
*Myrothecium verrucaria, 64, 70, *71*

Nectria, 106
*Nectria cinnabarina, 100, *101*, 106
*Nectria haematococca, 104, *105*, 106
Nematospora, 75
*Nematospora gossypii, 75, 76, *77*

Oedocephalum lineatum, 156, 160, *161*
*Oidium, 84, 86, *87*
Orbimyces, 64
*Orbimyces spectabilis, 70, *71*
Ostracoderma epigaeum, 113, 116, *117*
Otidea, 113
*Otidea concinna, 114, *115*
Ovulariopsis, 84

Panus, 146
*Panus rudis, 150, *151*
Papulaspora, 63, 68, *69*
*Parmelia, 176, *177*
Peltigera, 174
*Peltigera aphthosa, 174, 176, *177*
*Penicillium, 12, 13, 63, 68, *69*, 80
Penicillium claviforme, 63
Penicillium camemberti, 188
Penicillium chrysogenum, 188
Penicillium citrinum, 189
Penicillium crustosum, 189
Penicillium cyclopium, 189
*Penicillium expansum, 180, 182, *183*, 189
Penicillium islandicum, 189
Penicillium italicum, 180
Penicillium notatum, 188
Penicillium purpurogenum, 189
Penicillium roqueforti, 188
Penicillium vermiculatum, 80
Peronospora, 45
Peronospora effusa, 45
Peronospora geranii, 45
Peronospora manshurica, 45
*Peronospora parasitica, *48*
*Pesotum ulmi, 76, *77*, 81
*Pestalotia, 65, 70, *71*
*Peziza, 112, 116, *117*
Peziza ostracoderma, 113, 116, *117*
*Peziza varia, 114, *115*

Phallus, 168
*Phallus impudicus, 168, *169*
*Phoma, 65, 70, *71*
Phomopsis, 102, 189
Phycomyces, 49
*Phycomyces nitens, 49, 54, *55*, 56, *57*, 184
*Phyllachora graminis, 89, 90, *91*
*Phyllactinia, 84, 90, *91*
Physarum cinereum, 28
*Physarum polycephalum, 28, 30, *31*, 32, *33*
*Physcia, 164, *165*, 174
Phytophthora, 5, 45
*Phytophthora infestans, 45, *48*
Phytophthora palmivora, 45
Pilobolus, 50, 179
*Pilobolus kleinii, 54, *55*
Pinus, 4
Pinus elliottii, 172
Pinus taeda, 172
*Pisolithus, 164, *165*, 167
*Pisolithus tinctorius, 167, 168, *169*, 172
Plasmodiophora, 43
*Plasmodiophora brassicae, 32, *33*, 43
Plasmopara, 45
Plasmopara geranii, 45
*Plasmopara viticola, 45, *48*
Pleurotus, 146
*Pleurotus ostreatus, 146, 150, *151*
*Podosphaera, 84, 86, *87*
Polyporus, 156, 160, *161*
Polyporus berkeleyi, 156
Polyporus cinnabarinus, 156
Polyporus sulphureus, 156
Polyporus tulipifera, 157
Polyporus versicolor, 157, 160, *161*, 162, *163*
Polysphondylium, 23
*Polysphondylium violaceum, *25*
Poria, 157
Poria cocos, 157
*Poria incrassata, 157, 162, *163*
Protostelium, 23
*Protostelium mycophaga, 23, *25*
*Pseudocyphellaria anomala, 174, 176, *177*
Pseudopeziza, 118
*Pseudopeziza medicaginis, 116, *117*, 118
Psilocybe, 146
Psilocybe aztecorum, 146
*Psilocybe caerulescens, 146, 152, *153*
Psilocybe cubensis, 146
Psilocybe mexicana, 146
Puccinia, 136, 137
*Puccinia graminis f. sp. tritici, 134, *135*, 136

Pycnoporus cinnabarinus, 156
Pythium, 44
Pythium aphanidermatum, 44
*Pythium dissotocum, 44, 46, *47*
*Pythium ultimum, 44, 46, *47*

Rhizidiomyces, 42
Rhizophydium, 35
*Rhizophydium pollinis-pini, 35, 38, *39*
Rhizophlyctis, 36, 38, *39*
Rhizopus, 50
Rhizopus arrhizus, 188
*Rhizopus nigricans, 50, 54, *55*, 56, *57*
Rhodotorula, 62
*Rhodotorula rubra, 62, 68, *69*
Rhytisma, 122
*Rhytisma acerinum, 116, *117*, 122
Rozella, 36
*Rozella allomycis, 36, 38, *39*
Russula, 106, 144
*Russula brevipes, 150, *151*
*Russula emetica, 144, 160, *161*

Saccharomyces, 75
*Saccharomyces cerevisiae, 75, 76, *77*, 184
Saccharomyces cerevisiae var. ellipsoideus, 188
Saccharomyces kluyveri, 75
Saprolegnia, 44
*Saprolegnia ferax, 44, 46, *47*
Sarcoscypha, 112
*Sarcoscypha coccinea, 112, 114, *115*
Schizophyllum, 154
*Schizophyllum commune, 154, 158, *159*
Schizosaccharomyces, 75
*Schizosaccharomyces octosporus, 75, 76, 77
Schleroderma, 167
*Scleroderma geaster, 167, 168, *169*
Sclerotinia fructicola, 118
Sclerotinia sclerotiorum, 189
Sclerotium, 63
*Sclerotium bataticola, 63, 68, *69*
*Sclerotium rolfsii, 63, 68, *69*
Septobasidium, 141
*Septobasidium burtii, 138, *139*
*Septoria, 65, 70, *71*
*Septoria apii, 65, 70, *71*
*Smittium sp., 53, 60, *61*
Sordaria, 94
*Sordaria fimicola, 94, 96, *97*, 100, *101*, 184, 186, *187*
Sparassis, 155
*Sparassis crispa, 155, 158, *159*

Spathularia, 119
*Spathularia flavida, 119, 120, *121*
Sphacelia segetum, 108
Sphaceloma, 127
Spilocaea pomi, 130
*Spiniger meineckellus, 156, 160, *161*
*Sporobolomyces, 62, 68, *69*
Sporormiella, 130
*Sporormiella australis, 128, *129*, 130
Starkeyomyces, 65
Steccherinum, 155
*Steccherinum septentrionale, 155, 162, *163*
Stegophora, 102
*Stegophora ulmea, 102, 104, *105*
*Stemonitis, 29, 32, *33*
*Stemonitis nigrescens, 30, *31*
Stereum, 155, 162, *163*
Stereum frustulosum, 155
Stereum gausapatum, 155
Strobilomyces, 144
*Strobilomyces floccopus, 144, 152, *153*
Syncephalastrum, 50
*Syncephalastrum racemosum, 50, 58, *59*, 60, *61*
*Synchytrium endobioticum, 35, 38, *39*

Talaromyces, 80
*Talaromyces flavus, 76, *77*, 80
Talaromyces vermiculatus, 80
Taphrina, 75
Taphrina communis, 75
*Taphrina deformans, 75, 76, 77
Thamnidium, 50
*Thamnidium elegans, 50, 54, *55*, 60, *61*
Thelephora terrestris, 172
*Thielavia terricola, 90, *91*, 94
Thielaviopsis, 64
*Thielaviopsis basicola, 64, 70, *71*
Tilletia, 138
Tilletia caries, 138
Tremella, 140
*Tremella mesenterica, 138, *139*, 140
Trichoderma, 106, 184, 186, *187*
*Trichophyton mentagrophytes, 70, *71*, 72
*Tuber, 116, *117*, 120, *121*, 123
*Tubercularia vulgaris, 100, *101*, 106

Umbilicaria, 174
Uncinula, 84
*Uncinula macrospora, 86, *87*
Urnula, 112
*Urnula craterium, 112, 114, *115*, 116, *117*
Uromyces, 137

*Uromyces phaseoli, 134, *135*, 137
*Usnea, 176, *177*
Usnea salacina, 174
Ustilago, 138
Ustilago avenae, 138
*Ustilago maydis, 134, *135*, 138, *139*
Ustilago nuda, 134, *135*, 138

Vaccinium, 65
Venturia, 130

*Venturia inaequalis, 128, *129*, 130
Vibrissea, 120
*Vibrissea truncorum, 120, *121*

Xylaria, 98
*Xylaria hypoxylon, 96, *97*, 98, 100, *101*

Zygorhynchus, 50
*Zygorhynchus vuilleminii, 50, 56, *57*